The GEM KINGDOM

_by_

Paul E. Desautels

Choosing your paints

The kind of paint you choose depends on the type of painting you want to do, how much money you want to spend and whether you want to carry your equipment around or keep it in one place. There is no reason why you should restrict yourself to one kind, but remember that each type will require slightly different equipment to go with it.

All paints are made of a coloured powder, called pigment, mixed with something to bind it together. This "binder" is different for each type of paint and it is this which gives the different types of paint their individual qualities.

Oil paints

Oil paint is made of pigment bound with oil. It is sticky and slow to dry. This makes it rather messy to use and storing and carrying paintings you are working on may be difficult. Its slow-drying quality means that you can correct mistakes and make alterations quite easily and it can give a greater variety of effect than any other type of paint. It is equally good for painting detail and broad sweeps of colour and can be used thickly or in thin transparent washes.

Water-colours

Water-colour paint is made of pigment mixed with gum and diluted with water. It is more liquid than oil paint and has to be applied more thinly. It dries quite quickly, which makes it difficult to correct mistakes.

It can be used in many ways but is best for small scale work. You need less extra equipment than for oil painting and it can be very light and compact. For this reason it is a good paint to use when painting out-of-doors.

Gouache

Gouache (pronounced "goo-ash"), like water-colour, is pigment mixed with gum, but it also has white pigment added to it to make it opaque (non-transparent). It comes in various forms – powder and poster paints are both types of gouache. It dries quite quickly and you can paint over it to correct mistakes. It can be used thickly, like oil paint, or diluted and used almost like water-colour. If you use it thickly it has a tendency to crack and flake.

Acrylics

Acrylic paints are a fairly recent development. They are made from pigment mixed with a chemical binder. The range of colours available in acrylics is wider than in other types of paint.

Used by themselves, or with water, the paints dry very fast and are difficult to blend or alter. You can also mix them with a special "medium" to slow down the drying process, so you can use them like oils.

The cost of the paint is roughly the same as oil paints and water-colour.

Cost and quality

The price of paints varies enormously, according to the type (oils, water-colours, acrylics or gouache), the make, the quality and the colour.

On the whole the more paints cost the better quality they are. This means that they have been mixed better and are less likely to be grainy or lumpy. The colours may be slightly richer and less likely to fade and the paint is less likely to deteriorate with age. For a beginner the important thing is to experiment, and, as a rule, it is not worth buying the more expensive paints.

The cheapest type of paint to use is gouache, especially if you use poster or powder colours. Oils, water-colours and acrylics are roughly comparable in price but water-colours need less extra equipment and tend to go further as you would probably use them more thinly than oils and acrylics.

Oil painting equipment

Some paint manufacturers sell special boxes, complete with brushes, paints, palettes and the other materials you need for oil paintings. But, as a beginner, you will probably find it cheaper and more practical to buy all your materials separately.

If you are going to paint outside you will need something to keep your paints in, but you can use anything from a plastic bag to a purpose-built box, depending on how much you want to spend. An ordinary tool box is quite useful for holding paints and other small pieces of equipment.

Paints

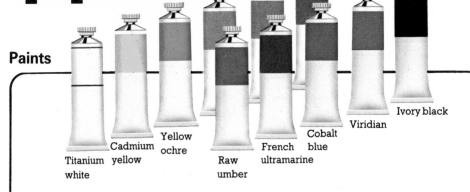

Titanium white · Cadmium yellow · Yellow ochre · Cadmium red · Alizarin crimson · Indian red · Raw umber · French ultramarine · Cobalt blue · Viridian · Ivory black

Oil paints are sold in tubes and most manufacturers produce two grades. The ones labelled "Artists'" are the better quality, in that the colours are stronger and do not fade with age. The lower grade, sometimes labelled "Students'" are cheaper and just as good for beginners.

Start with a fairly small range of colours, because this makes you experiment with mixing colours together to get the one you want. The 11 colours shown above are a good selection to start with.

Brushes

Brushes vary in size, shape and texture. They are made either from animal hair, or from synthetic material, such as nylon. The hair ones are more expensive, but if you take good care of them they will last longer. Soft-haired brushes are made from sable (a small arctic animal) or squirrel hair. Stiff-haired brushes are made from bristle (hog's hair). Try to start by having at least three brushes, one soft and two stiff ones of different shapes and sizes.

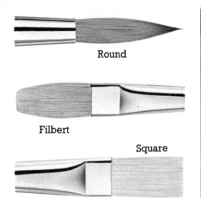

Round

Filbert

Square

Brushes come in three basic shapes: round, filbert and square. The round ones give the greatest range of marks.

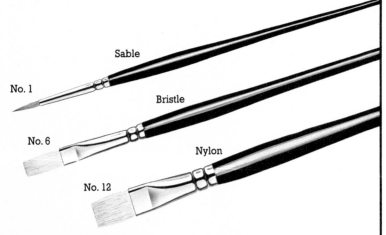

Sable
No. 1

Bristle
No. 6

Nylon
No. 12

Brushes come in a range of sizes, numbered from one to 12. Number one it the smallest size. Small brushes are for detailed work and thick ones are for covering large areas as quickly and smoothly as possible.

Palettes and palette knives

Palettes and palette knives are used for mixing colours before you put the paint on your picture. You can use anything you like as a palette – a piece of formica, a tin lid, an old plate, or board covered in tin foil or cling film – as long as the paint does not sink into its surface and it is plain-coloured.

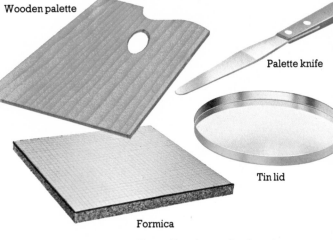

Ready-made palette. These also come in oval or kidney shapes.

Wooden palette

Palette knife

Tin lid

Formica

Piece of formica stuck to board

If you use a piece of wood, first smooth it with sandpaper and then rub some linseed oil into it, to stop the paint sinking in. Before buying a ready-made palette, check that it is comfortable to hold. You can also buy disposable palettes made of oil-proof paper.

There are many different types of palette knife. Choose one that feels comfortable to hold.

Mediums

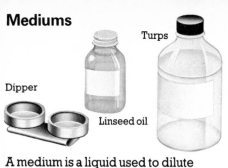

Dipper

Turps

Linseed oil

A medium is a liquid used to dilute oil paint in order to give it the consistency you want. Sometimes you can use the paint straight from the tube, but often you will want to thin it. For this you will need turps or white spirit, and linseed oil. You can use an egg cup or something similar to hold the medium, or buy a little cup called a dipper, which clips on to the edge of a palette.

Easels

You can adjust the height of your canvas or board, which is held between these two points.

Radial easels can be folded down and tilted backwards or forwards.

It is not necessary to have an easel. Indoors you can usually find a chair or a table where you can prop your painting at a comfortable height for you to work at. If you do decide to buy an easel, make sure you get one that is fairly solid. Studio easels are sturdy and adjustable, but they are not collapsible, they take up a lot of space and they are expensive. A radial easel, shown above, is a good alternative.

Painting surfaces

Canvas is probably the best surface to paint on, but it is also expensive. You can buy oil painting paper or specially prepared board. The boards either have real canvas stuck to them, or a cheaper material that looks like canvas. At first it is probably better to make your own oil painting boards.

Hardboard. You can usually have it cut to the size you want in the shop where you buy it.

Prepared canvas

Sandpaper

Primer

Household paint brush

If you cut your own hardboard to size, use a tenon saw, like this one.

To make your own board you need a piece of hardboard, some fairly fine sandpaper, some hardboard primer or white emulsion paint and a household paint brush. Lightly sandpaper the smooth side of the hardboard and then apply the primer or emulsion with the brush.

Cleaning materials

Rag

Soap

White spirit

It is important to clean your palette and brushes after you have used them. To do this you will need white spirit, which you can buy in a hardware store, a clean rag and a bar of soap. For more information about how to clean your equipment, see page 7.

Check list of equipment

About 11 tubes of paint
At least 3 brushes
A palette/piece of formica/tin lid/ old plate
A bottle of linseed oil
A bottle of turps
A board, canvas or piece of oil-painting paper
An easel, chair or table
A bottle of white spirit
A rag
A bar of soap

How to use oil paints

You may want to do an oil painting quite quickly, using fairly thin paint, or you may want to build up your picture more slowly over a fairly long period of time. If so, try to choose a place to work where you can leave your equipment set up when you take a break.

Oil painting can be rather messy, so make sure you cover your work area with newspaper, and wear an overall or old shirt to protect your clothes.

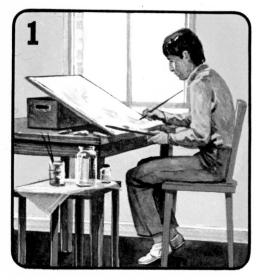

You can put a little turps and linseed oil in dippers attached to your palette.

Make sure that you set up your easel somewhere where you have good light and can see your subject comfortably if you are painting from life. If you have not got an easel, prop up your canvas or board on a table or chair at a comfortable height for you to work at. It is useful to have a small table beside you, for holding your equipment.

There are many different ways of arranging the paints on your palette, but if you always set them out in the same way, you will soon know, almost without thinking, where each colour is. It is a good idea to keep the lighter colours on one side of the palette and the darker ones on the other, as shown here.

Put your full selection of colours out on the palette. You never know which ones you will need. They dry slowly, so you do not need to worry that they will go hard.

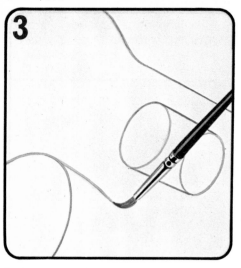

Most people find it helpful to begin an oil painting by making an outline drawing, or at least sketching in roughly where each object will be in the picture. The best thing to use for this is a small brush and paint thinned with turpentine. Use brown or blue, rather than black, which takes longer to dry.

If you want to mix your paints to match the colours to your subject, first choose a particular patch of colour and select the colour on your palette that is most like it. Use your palette knife to put some of this colour in the middle of your palette. Decide whether you need to make it lighter or darker and start mixing in other colours, one at a time, until you arrive at a colour as close as possible to the selected part of your subject. Try to keep the mix as simple as possible as overmixing can make the colours look dull and muddy. Always use your palette knife for heavy mixing. If you use your brushes they will get clogged up with paint.

Using thinners

Dipper filled with turps.

Take care not to muddy your turps by stirring it with your brush.

To make the paint easier to apply you will need to mix a few drops of turps with it. In the early stages of a painting particularly, the paint should be fairly thin. In the later stages you can mix a little linseed oil with your turps. This helps to keep the colours bright after the painting has dried.

Making corrections

If you make a mistake while you are painting, or you want to change something, it can be done quite easily. With your palette knife, scrape the paint from the area you want to re-paint. Dip a rag in turpentine and rub the area until it looks clean. You can then re-paint it. For small mistakes, simply paint on top of them, using slightly thicker paint.

Applying the paint

Oil paint can be applied in many different ways. Below you can see two of the different effects you can achieve. You will discover many more effects as you get more experienced. At first it is better not to try to get a very detailed or finished look. Do not load your brush with too much paint and try to use just the tip. Scrubbing the surface to remove all the paint will damage the brush.

◀ One method is to apply quite pure, unmixed colour in small strokes, without merging or blending them. The crisp brush strokes make the painting lively. From a distance the effect becomes more subtle as the colours are blended by the eye. This is quite a slow way of working, so it is better to keep it for small pictures.

Another method is to use larger areas ▶ of unbroken colour. The shades need to be more carefully mixed on the palette and the paint can be applied with larger brushes. This sort of painting needs to be more carefully worked out beforehand.

Clearing up

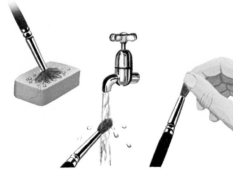

When you have finished painting use a palette knife and some newspaper to scrape the paint off your palette, and then wipe it with a rag dipped in white spirit. A little linseed oil can be rubbed into wooden palettes to protect them. Never allow paint to harden on your palette. You can leave the colours round the edge for up to about eight hours before they will dry out, but the paint that has been thinned and mixed in the centre will dry more quickly. Wipe your brushes on a rag and then rinse them in white spirit. Wash and rinse them well with soap and water. Dry them well and smooth them back into shape with your fingers.

Storing and varnishing

Make sure all the caps are screwed tightly on to your tubes of paint, otherwise the paint will go dry. Store your brushes somewhere where they will not get dusty and lie them flat so they keep their shape.

Store your painting away from direct heat. When the paint has completely dried, you can varnish it to protect it from dust and give it a glistening effect, although some people prefer unvarnished paintings. Allow several months for the paint to dry – thick paint may take up to a year. You can buy varnish in bottles or spray cans with instructions on how to use it printed on them.

Water-colour equipment

The equipment you need for painting with water-colours is simpler and lighter than that needed for oil painting. Good quality paints and brushes are worth the extra cost as they give better results and are easier to use.

Brushes

Squirrel

Sable

Most people prefer soft brushes for water-colour painting. Choose a range of different sizes. A good basic selection might be: one large wash brush made of squirrel hair, a medium round sable brush and one small squirrel or sable brush.

Easels

Sketching easels are lightweight and easy to carry about with you. You can also use a stool and sit with your board propped up on your knees. A camping stool is light and easy to carry. Some people prefer to stand with one leg propped up on something while resting their drawing board on their knee.

Paints

You can buy water-colour paints either in tubes, or in little blocks called pans. The pans come in two sizes – half-pans and whole pans. The paint in the tubes is easier to mix with water to make a light colour wash for a large area of paper, but it also tends to harden in the tubes after a time.

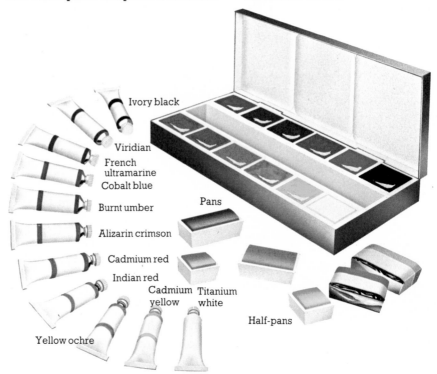

Ivory black

Viridian

French ultramarine

Cobalt blue

Burnt umber

Pans

Alizarin crimson

Cadmium red

Indian red

Cadmium yellow

Titanium white

Half-pans

Yellow ochre

Water-colour pans and tubes are often sold in boxes. You can also buy empty boxes to fill with your own selection of pans and tubes. It is useful to keep your paints in a paint box, as it also provides a palette on which to mix your paints.

You will need about 11 colours to begin with. The ones listed above, which are the same selection as for oil paints, would provide you with a good range to start with.

Palettes

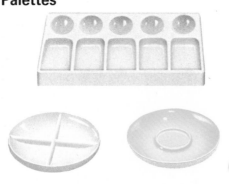

Paint boxes usually have one flat surface for mixing thicker paint, and one deeper palette for watery mixes. If you do not have a paint box, you will need to find a mixing surface. A well palette, or a collection of small pots, allows you to mix in plenty of water without the colours running together.

Other equipment

You need plenty of clean water. Change the water in your jar as soon as it starts to get dirty. Rags and sponges are useful for mopping up paint that runs, as well as for cleaning your brushes. You can also use sponges to apply paint.

Paper

Art shops sell a wide variety of special water-colour papers. These vary a great deal in texture, thickness, colour and price. You can buy them in single sheets or in sketch books and pads, which are much better value.

You need not use special water-colour paper. Any paper can be used, as long as it is not too thin and shiny. Sugar paper is particularly cheap. Unless you use a very thick type, you should stretch your paper, as shown below, to stop it crinkling when you paint on it.

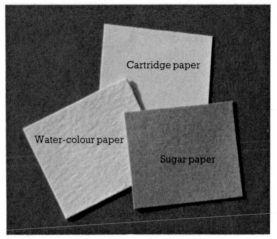

Cartridge paper

Water-colour paper

Sugar paper

Water-colour paint appears at its most transparent and delicate on white paper. Coloured paper can sometimes be very effective, but your paint may seem to lose its strength and purity.

1 How to stretch paper

2

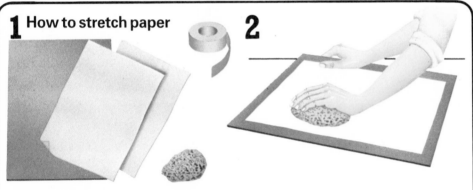

Place one piece of paper in the centre of the hardboard. Dip the sponge in the bowl of water and then squeeze it. Wet the paper thoroughly with the damp sponge.

3

4

Stick gummed paper along full length of each edge.

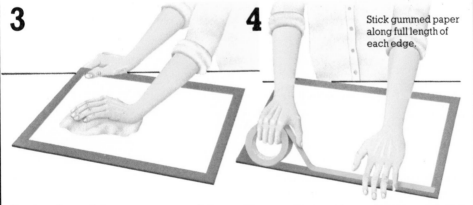

Put the piece of dry paper on top of the wet one, to avoid damaging the wet surface. Smooth it flat with a clean cloth, working from the centre outwards.

Remove the top piece of paper and stick the bottom one down with the gummed paper. Leave it to dry away from direct heat. Make sure it is dry before using it.

Drawing boards

You need some sort of hard surface to attach your paper to while you are painting, unless you are using a sketchbook. You can buy ready-made drawing boards, or improvise with chipboard or plywood. Use clips, drawing pins or masking tape to attach your paper to the board.

Check list of equipment

About 11 tubes or pans of paint
3 brushes
A drawing board + clips or tape
A well palette/paint box with
    palette/saucers/egg cups
A sponge
A rag
A jar of water
A sheet of paper or sketch book

How to use water-colours

Water-colours are transparent. The difference between using them and other kinds of paint is that you use the whiteness of the paper to show through your colours, instead of using white paint.

You can use water-colours to paint any subject, but they are specially good for paintings which show the effects of light, or conjure up a particular mood or atmosphere. In the past they were often used for landscape paintings.

With water-colours it is difficult to change things once they have gone wrong, so you will probably find that you make several disastrous attempts before you get good results.

1

Because water-colour equipment is light and compact, it is easy to carry around and particularly good for painting out-of-doors. Make sure you have a good supply of clean water and that the light on your paper is good but not glaring.

2

It is difficult to make alterations once you have applied paint to your paper. For this reason drawing usually plays an important part in water-colour painting. Use a pencil (HB or B) or pen and waterproof ink. The drawing will show through the paint to some extent. If you want your picture to have very strong outlines you can also use pen and ink on top of dried paint.

You can, of course, start painting without doing any preliminary drawing. You need practice to do this, so you may want to use a brush and thin paint mixture to draw in a few guiding marks.

3

If you are using tubes, do not squeeze out the colours before you need them, as they dry out quite quickly on the palette.

It is important to keep your colours as fresh-looking as possible. Try not to mix them too much. Change the water in your jars regularly too, as dirty water will also make the colours look dull. Test your colours on a scrap of paper before you apply them to your painting. Wet colour often dries to give a lighter tone.

4

As a general rule, apply lighter colours first and when these have dried, gradually add the darker shades you need. Most people start by getting their colours much too dark. It is very difficult to lighten dark shades of water-colour once you have applied them, as the paint sinks into the paper and stains it.

You can use the white of the paper for any white objects and for highlights, but you can also use white paint.

Laying on a colour wash

If you want a large area of soft, flat colour, you can use a wash of thinly diluted paint.

To mix paint for a wash pour out enough water to cover the whole area to be painted. Add paint gradually until you have a strong enough shade.

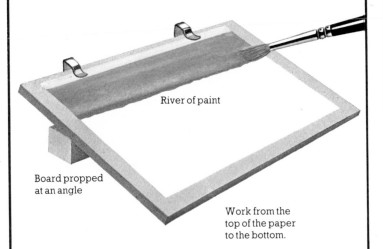

River of paint

Board propped at an angle

Work from the top of the paper to the bottom.

Prop your board at a slight angle. Fill a large brush with watery paint and apply it with smooth strokes in a band across the top of the paper. As you paint a river of paint will form along the bottom of the band. Pick this up with your brush and use it to paint the next band of paint. Whenever the river of paint runs dry, recharge your brush and carry on from where you left off. When you have painted from top to bottom, dry your brush and then collect up any wet paint still left at the bottom.

Sketchbooks

Water-colours are very useful for making quick sketches. Use a sketchbook to record your impressions of things you see – impressions of light and shadow, shapes and colour combinations that interest you. This is a good way of teaching yourself to observe things accurately and to spot interesting subjects.

A sketchbook is also useful for small, detailed paintings, such as studies of fruit and other natural or man-made objects which have an interesting shape, texture or colour.

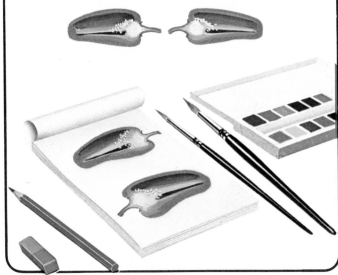

Correcting mistakes

If you do make a mistake it is sometimes possible to dab the paint out with a damp sponge or tissue. Be careful not to damage the surface of your paper. Let the paper dry before painting on it again, or the paint may sink into the paper, leaving a blotchy effect.

Cleaning up

Wash your palette when you have finished painting, or, if you are using a box, wipe it clean with a damp rag. If you are using pans and they have become dirty, wipe them over also with a damp rag. If paint is left to harden on them, it will stain them.

Wash your brushes in cold water and reshape the hairs to a fine point with your fingers.

Storing

If you use pans and do not use your paints very often, put a small piece of damp sponge in the corner of the box to prevent them from drying out. If you use tubes of paint, make sure that the caps are tightly screwed on.

Keep your brushes in a tube or box to prevent the moths getting at them. Keep this flat or upright with the hairs pointing upwards so that they keep their shape.

Store your paintings flat and away from direct heat.

Gouache paints

Gouache, like water-colour, is a water-based paint, but it has white pigment added to it. This makes it thicker than water-colour and opaque, so that the paper does not show through the paint.

White   Lemon yellow   Brilliant yellow   Scarlet

Crimson   Magenta   Brilliant green   Viridian green

Coeruleum blue   Cobalt blue   Ultramarine blue   Spectrum violet

Yellow ochre   Raw umber   Burnt sienna   Black

Gouache paint does not mix to produce such clear colours as you can get with oil paints and water-colours. For this reason it is a good idea to have a few more colours than you would need for these other methods. The names of the colours are sometimes different from similar colours in oils or water-colours. The range shown here would be a good selection to start with.

Paints

Gouache comes in various different forms.

Poster paints come in paste form in jars.

Some manufacturers produce poster paints in large containers. The paint name varies according to the brand.

Powder paints come in tins and are cheap. To mix them put some powder in a palette and add small amounts of water at a time. It is difficult to get a smooth mix by adding powder to water.

Designers' gouache colours come in tubes, jars and pans and are the best quality and most expensive of the gouache paints.

Pans

Brushes

You can use soft water-colour brushes, or bristle brushes, depending on the effect you want to create. Try to have a good range of shapes and sizes.

Painting surfaces

If you paint with gouache, you can use a greater variety of surfaces to paint on than you can for any other type of paint. Most types of paper or card are perfectly all right to use.

Other equipment

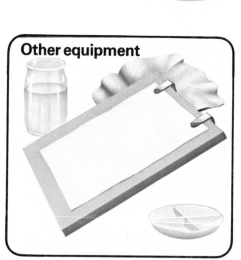

As for water-colour painting, you will need a jar of water, a rag, a well palette or several small dishes and a drawing board with clips or tape.

Using gouache

Gouache is an ideal paint to experiment with because it is the most economical type of paint and the most straightforward to use. Here are some suggestions:

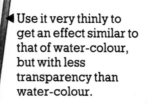

◀ Use it very thinly to get an effect similar to that of water-colour, but with less transparency than water-colour.

◀ Try doing very bold large-scale paintings. Use poster paints and large, cheap pieces of paper, stuck together and pinned to a wall or floor.

Detailed work using designer's gouache

◀ Large areas of flat colour

Use the paint straight from the pot or tube to experiment with colours and their effect on each other. Do large areas of flat colour or cover the surface with brush strokes without blending the colours.

▲ If you want to do detailed work with gouache, it is best to use designer's colour and fine grained paper. If you are trying to match colours to a subject, remember to test them on a scrap of paper first.

Check list of equipment

About 16 jars, tubes or pans of paint
A selection of brushes
A sheet of card or paper
2 jars of clean water
A rag
A palette/saucers/plates
A drawing board
Clips or tape for attaching paper to drawing board
Scraps of paper for testing colours

Cleaning and storing

When you have finished painting wash your palette and brushes in plenty of cold water. Make sure that the lids and caps are screwed tightly on to your tubes or jars.

Store all pots and tubes of paint away from direct heat or sunlight, as they may dry out or even change colour slightly.

The paintings should also be kept away from direct heat as gouache colours do sometimes fade. Once they are completely dry, store paintings by lying them flat, as the paint has a tendency to crack, especially if it has been thickly applied. If you store several paintings on top of each other it is a good idea to put a piece of tissue paper between each layer.

Acrylic paints

Acrylic paint was invented fairly recently. It is a very versatile type of paint and the equipment you need will depend on how you choose to use it.

Painting with acrylics requires less basic preparation than with other types of paint. They dry very quickly, will not crack or fade and are waterproof.

Paints

You can buy acrylics in either jars, or metal or plastic tubes, depending on the make you buy. The paint in jars is thinner than the paint in tubes.

Mediums

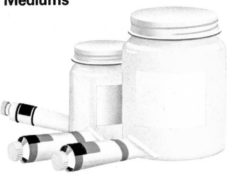

You can buy various acrylic "mediums" which you can mix with the paint to give it a more matt or gloss effect. There is also retarder medium, which you mix with the paint to slow down the drying process. You can also use water to thin them down.

Colour range

Acrylic colours vary according to which make you buy, but there is generally a wide range. You can get some much brighter and more eye-catching colours than are available in other types of paint. You can also get the more traditional colours although these tend to lack the richness of the same colours in oil paints. For a good colour range see page 4. Below you can see some of the newer colours that are available.

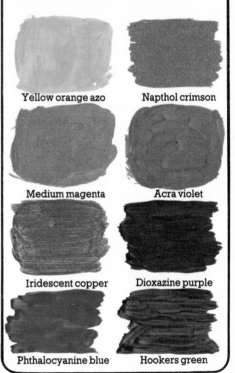

Yellow orange azo   Napthol crimson

Medium magenta   Acra violet

Iridescent copper   Dioxazine purple

Phthalocyanine blue   Hookers green

Other equipment

Any water-resistant surface is suitable for use as a palette. Do not use a wooden palette, as they are very hard to clean. When you want to clean your palette soak it in water and the paint will lift off in a layer, rather than dissolving.

You can use a wide range of surfaces such as canvas, board, card or most types of paper. If you use canvas or board you should prepare it by painting it with special acrylic primer. Whatever you paint on, make sure that it is not greasy. You cannot mix acrylics with oil.

Use soft or hard brushes depending on the effect you want. You must keep your brushes in water while you are painting and wash them in clean water immediately after you have finished. If paint dries on them it is extremely hard to get out.

Check list of equipment

A range of about 12 paints
A selection of mediums (retarder, matt, gloss)
A selection of synthetic or bristle brushes
Canvas, board, paper or card
2 jars of clean water (one for cleaning brushes in)
A rag
A water-resistant palette (china or plastic), piece of formica or something similar
A drawing board and clips (if you are using paper)

Using acrylics

Acrylic paint is not suitable for certain types of painting. Its quick-drying qualities mean that it is not easy to make adjustments. It is best used for paintings you have planned out quite carefully before applying any paint. It is less easy to paint subjects like landscapes or portraits which need constant adjustment.

You can use acrylics in any of the ways shown below.

Some makes of acrylics are transparent, but you can mix them with white medium to produce a paint which can be used like gouache. Unlike gouache this paint will not crack or flake off.

Acrylic tube paints are as thick as oil paints and can be used in the same way as oils. If you use them like this, you may want to use a retarder medium to slow down the drying process and give them an easier texture to work with.

If you mix water with acrylics, you can use them as though they were water-colours and produce transparent washes. These can only be blended while the paint is still wet because once it dries it is waterproof. Once a wash is dry you can overpaint without fear of it running.

Hard-edge painting

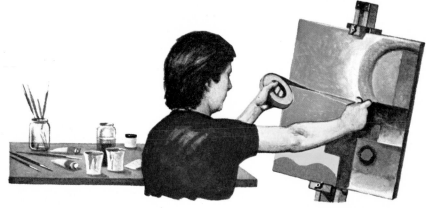

Acrylic paint is particularly suitable for painting flat, smooth areas of colour with straight edges. You can use masking tape to help you get straight edges. Stick the tape down firmly along the edges of the area you want to cover, and paint up to and over it. When the paint has dried, tear off the tape. You can also stick masking tape over dry paint to help you paint up to a line without fear of the colours merging together.

Staining

You can create interesting effects by staining unprimed canvas or fabric. To do this paint or sponge on layers of water-thinned acrylic colour. You can do a complete painting like this or just a background.

Collage

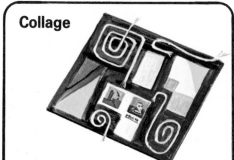

Acrylics are well suited for collage because the mediums are adhesive. Use thick paint and press other materials, such as string, cork or gravel, into it. Use hardboard or stiff card to paint on, because paper will buckle.

Airbrush

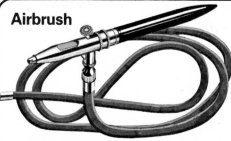

Airbrushes work by using compressed air to blow paint on to a surface. With them you can produce fine lines, softly graded tones and solid areas of even colour. They vary enormously in price and even the cheapest ones are fairly expensive. You can use most types of paint in them. Acrylics are often used because they dry quickly. They must be diluted with water and the airbrush must be cleaned immediately after use.

Choosing what to paint

Choosing what to paint can be surprisingly difficult at first, because there is such a wide choice. However, certain practical considerations can often help you to make up your mind.

First, you need to decide whether you want to paint something from life (something you can actually see in front of you), from your imagination or from a mixture of the two. To begin with you will probably find it easier to paint from life.

You also need to decide where you are going to do your painting, in particular, whether inside or outside. Think about the light you will need and the weather conditions.

Your choice may be influenced by the type of paint you are going to use. Certain types of paint are better suited to certain types of painting*. If you have a choice of paints, you may choose your subject and then decide what paint to use.

A still life is a good subject to begin with, because you have time to study your subject without it changing. Flowers, food and ordinary household objects are good subjects for this kind of painting.

A view of a room can make a good picture. Think about the colours and shapes in your picture and experiment with different views until you find the one you like most. At first keep it quite simple.

If you want to paint a portrait, you will have to find someone with enough time to sit for you. If this is difficult, you can try using a mirror so you can paint yourself.

To paint a group of people you do not need to get everyone in the group to pose together. Choose a characteristic scene and build the picture up slowly, over a period of time.

If you want to paint an action shot, you will need to use photographs, sketches from life and other pictures to help you. It takes practice to capture a brief moment of action without making it look rather frozen and artificial.

Your choice of outdoor views and scenes to paint will be influenced by where you live and by what you can see from your windows. Remember that the light on your subject will change according to the time of day.

Painting completely from imagination is surprisingly difficult, although very exciting. You will probably need to use sketches from life or other pictures to help you, or to paint certain parts of the picture from life.

There are a number of different ways of getting yourself started on a picture that does not show realistic objects. You can start simply by putting down patches of colour and building them into patterns.

For more information see the pages on paints and how to use them.

Planning your paintings

When you have decided what to paint you need to plan where the things you have chosen will go in your picture. This is partly so you can be sure of getting in everything you want, but also because the success of your picture depends very much on how you have arranged the colours and shapes. The way you arrange your picture is called the composition.

When you paint from life, your composition depends a great deal on choosing a good position from which to view your subject. When you paint a picture from imagination you can arrange the same subject in any number of different ways. The important thing is to arrange the objects so that they look good together. Here are some hints to help you to do this.

One of the first things you need to decide is what size and shape your picture will be. If you are going to use paper, you may decide that the paper you have is the right size and shape, in which case you can work right up to the edges. Alternatively, you draw a square or rectangle on your paper and use this to mark the edges of your picture.

The first stage in planning your painting is to see all the objects in terms of outline shapes. Put in the larger shapes first and add the details later. It may help to do several quick sketches of different arrangements and then choose the one you like best. Stand back from your composition to see how well it works and be prepared to make adjustments.

Try to use a range of sizes for the objects in your pictures. If all your objects are roughly the same size, as in the picture above, it usually looks rather uninteresting.

Here the dancers are different sizes because some are further away than others. If you make objects overlap with each other it also adds interest, as well as helping to give an impression of distance.

Think about the balance of colour and tone* in your picture. Avoid having all the brightest colours on one side, as in the picture above, and try to keep a balance between light and dark tones.

A feeling of movement or energy can be emphasized by using diagonally placed shapes. Vertical lines give a more static effect and horizontal lines tend to give a feeling of calm.

Viewfinders

A viewfinder can be very useful when planning a painting. You can make one quite easily by cutting a window in a piece of card. By looking through it you can frame just as much as you want of the scene in front of you, in the same way that a photographer does when he looks through the viewfinder of a camera. Unlike a photographer, though, an artist can adjust and rearrange what he sees, in order to improve his picture.

A quick way of making it easier to select what you will paint can be done by closing one eye and with the other looking through the rough square made by joining your thumb and forefinger together.

See pages 20 and 21 for more information about colour and tone

Making sketches

You can start a painting by sketching in just a few rough guidelines, or even without drawing anything at all. Sometimes, though, you may want to do a more detailed drawing before you start.

You can draw your subject and then paint on top of your drawing or do your drawing on a separate piece of paper and use it as a reference to help you with your painting. Whichever of these approaches you take, drawing will help you to understand your subject better and give you ideas about how you want to paint it. Here are a few guidelines on how to draw.

Drawing materials

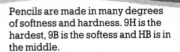

Conté crayons come in either sticks or pencil form. They are available in four colours – black, white, sepia and sanguine.

The best type of eraser to use is the squashy type known as a putty eraser. If it gets dirty, clean it by rubbing it on a piece of scrap paper.

Charcoal is sold in three sizes – thin, medium and extra large. It is best used for large, bold drawings. Do not use water-based paints to paint over it as it will dirty the colours.

Pencils are made in many degrees of softness and hardness. 9H is the hardest, 9B is the softess and HB is in the middle.

You can use thin paint and a brush to do a drawing, or ink with a pen or brush. If you want to do quite a detailed drawing it is best to choose something which makes marks that you can rub out. Each drawing instrument has its own particular feel.

Outlines

The simplest way of drawing is to use lines to show just the outlines, or most important features of the objects you see. This method is good for certain subjects but it does not give a feeling of solidity to the things you draw.

Directional lines

To show more detail you can use shorter, more closely-packed lines and dots. This helps to show the texture and feel of your subject as well as outline shapes.

Quick sketches

It is very good practice to do quick sketches using as few lines as possible to describe something. This trains you to look at something and to see immediately which features will conjure up the feeling you want to convey.

Shading

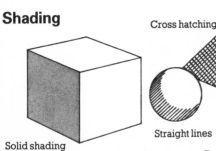

Solid shading

Cross hatching

Straight lines

Curved lines

Dots

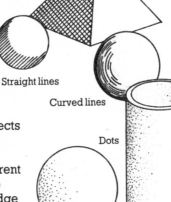

Shading helps to make objects look solid. There are many different ways of shading. Here you can see five different methods. If you want to use solid shadow you can smudge pencil, charcoal or conté lines with your finger.

Measurement and proportion

When you do a drawing you should measure the objects in your picture to see how they compare with each other in size, shape and position.

You can use your pencil for measuring and judging the exact position of things by holding it absolutely still at arm's length and closing one eye.

Use it to help you judge which things are on the same level, either horizontally or vertically. Here the man's hips are the same height as the top of the chair.

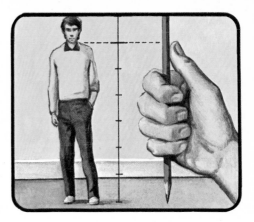

You can also use it to find out the proportion of one object to another. To work out the proportion of the man's head to his body, mark off the height of his head on the pencil with your thumb, and use this length to measure the body.

Perspective

Perspective is the art of giving an impression of solidness and depth to objects or scenes shown on a flat surface.

There are certain rules you need to follow if you want to achieve this. Some of these are shown below.

Things in the foreground appear to be bigger than distant things, even if they are really the same size.

Colours get less brilliant and the amount of detail you can see decreases as things get more distant.

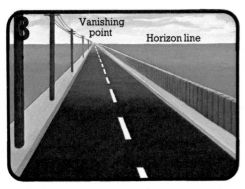

Lines that are in fact parallel appear to get closer together as they get further away and to meet on the line of the horizon. The point at which they appear to meet is known as the vanishing point.

By moving the horizon line up or down your picture, you can change the eye level from which the scene appears to be viewed. A high vanishing point makes you feel as though you are looking down on a scene, a low one that you are near ground level.

Getting lines at the correct angle

Use the system of measuring shown above to help you draw lines at the correct angles.

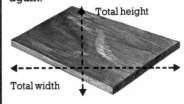

Hold your pencil at arm's length and slant it at exactly the same angle as the line you want to draw. Draw the line, then check by holding your pencil up again.

When drawing something like this table, it helps to get the shape exactly right if you measure the total width and total height.

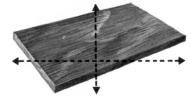

19

Using colour

Using colour is one of the most important aspects of painting. To paint something you need to look at it very carefully to notice all the different shades of colour, which you do not necessarily see at first glance. It will help you to get the effects you want if you understand a few basic things about colours and the way they affect each other.

The colour circle

There are three basic colours called primary colours. These are red, blue and yellow. You cannot make them by mixing other colours together, but you can make a wide range of other colours by mixing primary colours together.

By mixing any two primary colours together you make a secondary colour. Arranging the colours in a circle helps to show how colour mixing works. Each secondary colour is made by mixing together the colours on each side of it.

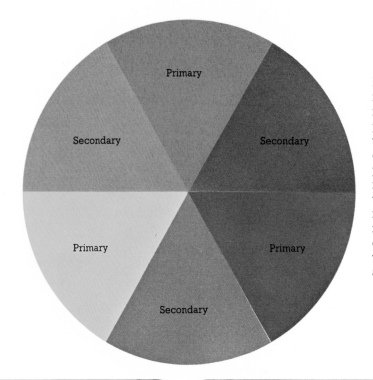

Primary

Secondary

Secondary

Primary

Primary

Secondary

In theory you should be able to mix most of the colours you might want from the three primary colours. In practice this is not possible as paint colours are not always very pure. This means that if you mix the paints on your palette too much, they will become muddy. It is a good idea to include at least two versions of each primary colour among the paints you choose, as well as black and white.

Effects of colours on each other

Colours appear to change according to the colours that surround them or are close to them. Each patch of colour that you paint will affect all the other colours in your picture. Some colours make each other look brighter, other colours kill each other.

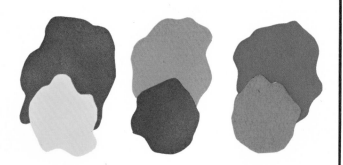

Colours opposite each other on the colour circle are called complementary colours. A colour always looks stronger when its complementary colour is near it. Careful use of complementaries can make a painting look very energetic and exciting.

Warm and cool colours

Some colours give a feeling of warmth, others have cool qualities. It is useful to bear this in mind when thinking about the overall mood you want to give to your painting.

 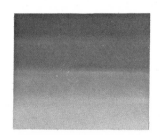

The warmer colours are reds, oranges, yellows and browns.

The cooler colours are blues, bluish greens and mauves.

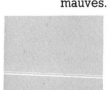 

Grey with blue   Grey   Grey with red

Small amounts of blue added to any colour will tend to cool it down.

Small amounts of red added to any colour will tend to make it look warmer.

Tone

The degree of lightness or darkness of a colour is called its tone. You can vary the tone of your paint colours simply by diluting them so that the colour is less concentrated. It can sometimes be difficult to judge the exact tone of a colour you see. A black and white photograph of your subject will show you clearly the differences in tone between the colours you see.

Black and white can be used for lightening and darkening other colours. However, you need to be careful because too much white can make a colour look chalky and too much black can deaden it.

On the other hand, when you want pure black or white in your painting, it is usually better to mix small amounts of another colour with them, to make the effect less harsh.

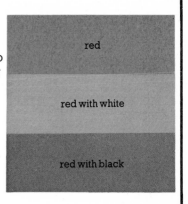

red

red with white

red with black

Light and shadow

Colour is determined by the sort of light in which you see it. Daylight often has a cold quality and certain colours, like blue and violet, stand out more strongly in it. Artificial light generally has a gentler and warmer effect.

It is tempting at first to use black to paint shadows and to show them much darker than they really are. The colours in shadows are usually very varied and rich, so use a range of colours for them.

Daylight

Artificial light

If you study a group of objects, you first of all notice the actual colour of each object. As you look harder and get more practice you will also see how colour is reflected from one object to another. The amount of reflection is determined by the surface textures of the various objects and the strength of the light which is causing the reflection.

Experiments with colour

Large piece of white paper

Small square of paper coloured with one of the primary or secondary colours.

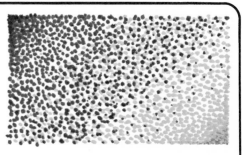

Arrange pieces of paper as shown above. Stare hard at the coloured square for several seconds. Remove the coloured square and keep looking at the white paper. This time you will see a square of the colour which is complementary to the colour of your first small square.

Cut out four small squares of red paper and place them in the middle of four different coloured backgrounds. Try using black, white, orange and green. Notice how the backgrounds affect the colour of the red. The green background will make the red look brightest.

Try painting different combinations of coloured dots to test how your eyes mix small amounts of colour together to make new colours. You will need to stand several metres away from the dots. Colour printing works like this – look closely at large colour posters to see the dots.

Still life painting

A still life is a painting that shows a single object or a group or arrangement of objects. This is probably the best kind of painting to tackle as a beginner. You can study your subject for as long as you like without it changing and you can set it up wherever you want and paint it at your leisure.

The traditional objects for still life paintings are things like fruit, flowers, dishes and jugs, but you can choose anything you like – things from around the house, tools and parts of machinery and natural objects like stones, shells and bits of wood, can all make interesting arrangements. Remember that if you choose anything perishable, like fruit or flowers, your painting time will be more limited than if you choose things that will not decay.

Accidental

Once you start looking for subject matter, you may discover an arrangement of objects that you want to use just as it is, such as a table after a meal, a chair draped with clothes, or this desk with lamp and books.

Composed

You may want to collect a group of objects together and arrange them yourself. You might try using a theme, such as the music theme used above, or choose things that make good colour combinations.

Arranging a still life

You need to think about the background and base against which you want to paint your objects. A backboard helps by cutting out distracting things behind them. A patterned or checked base covering can help you to work out the objects' exact positions.

To start with you will find it much easier if you restrict the number of objects in your still life arrangements to a maximum of five. Try to make the arrangement look balanced and let the objects overlap each other.

Choosing your viewpoint

Remember that you do not have to paint from straight in front of your subject. Try walking round it. You might discover a new and more interesting angle from which to paint it.

For variety try putting your objects on the floor and painting from above,

or put them on the edge of a table and sit on a low stool to paint them,

or arrange them on a window sill and use the view from the window as your background. Notice how shapes, such as the rim of the mug, change according to the eye level.

Lighting

The way the light falls on your arrangement makes a great difference to the shapes and colours you see. It is better to use artificial light, which remains constant, rather than daylight, which changes.

Use a fairly strong source of lighting and something that can be adjusted quite easily. A reading lamp or a spotlight are both good for this purpose. It is usually better if the light comes from the side so that you can see the shadows clearly.

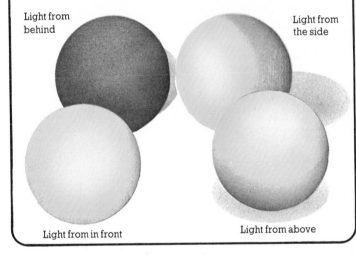

Light from behind

Light from the side

Light from in front

Light from above

Some ideas to try out

Try painting a selection of very small objects taken from a coat pocket, a bag or a sewing basket.

Try painting an arrangement using just one colour, in as many different shades as you need.

Set up a still life and paint only the background and the spaces between the objects.

Painting people

The human body can assume thousands of different positions. When you draw or paint people, one of the problems you will find is that they keep moving about, unless you can find somebody willing to be your model. So it is helpful to have a few guidelines on how to make people look realistic.

Studying photographs of people can be very helpful and making quick sketches is one of the best ways of learning how to paint them. Sketches can come in very useful when you are composing a larger picture.

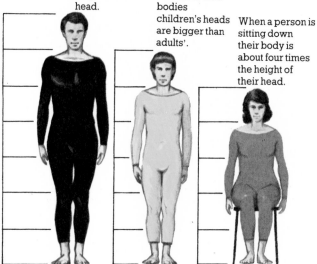

An adult's total height is normally about seven times the height of the head.

In proportion to the rest of their bodies children's heads are bigger than adults'.

When a person is sitting down their body is about four times the height of their head.

It is important to get the proportions of the different parts of the body right, in order to make figures look realistic. Use the system of measuring with a pencil, described on page 19, to help you work out how big the various parts of the body should be. Rules about proportion are only a rough guide. Each person's exact proportions are part of what makes them look different from everyone else.

To draw people in convincing positions it helps to think about how the human body keeps its balance. Every movement you make with any part of your body makes other parts of the body shift themselves to prevent you from losing your balance. Try noticing how this works by looking in a full-length mirror and moving your body into different positions.

It can be difficult to capture a feeling of movement when painting people. When people do things like running and jumping, the movements they make happen so quickly that you cannot really see what they are. If you can find photographs of people doing these things, they will help you to see the body position at different stages of the action.

To show people as solid shapes you need to paint in the highlights and shadows. Without shading they will tend to look like flat, cut-out figures.

Painting the folds in people's clothes can help you to make the figures look three-dimensional. Patterns and stripes can also help to show the form of the body.

When painting a group of people whose shapes overlap, it is easier to concentrate on the overall shape of the group, rather than treating each individual separately.

Portraits

If you want to paint a portrait, you first need to find someone to be your model. Choose a background and position to suit them and make sure they are comfortable. Do not expect your model to sit still for longer than about 15 minutes without a break. It might be best to let them read, sew or watch television while

you paint, to stop them getting bored.

You might want to start by doing a self-portrait, so that you can take as long as you like over your painting and work on it whenever you want. Set up a mirror where you can see yourself easily without moving your head too much.

Choosing your view

Three-quarter view

Profile

Full face

You have to decide how much of the body you want to include in your portrait – whether just the head and neck, or anything up to a full-length portrait. The position of the head is also important. A full face or three-quarter view is usually easier than doing a profile.

The head

Hair should be painted as areas of light and dark colour and not as a mass of lines.

The eyes are set about half-way down the head. People often think that they are higher, because the top of the head is hidden by hair.

Measure the distance between features within the face. Here the tops of the ears are roughly in line with the eyes.

Look carefully at the eye to notice how much of the coloured part you can see.

The whites of the eyes have quite a lot of colours in them.

Do not make the eyes and lips too dominant.

Think of the nose in terms of light and dark areas and do not be distracted by the outline.

Do not make the lips too bright. They are usually a pale reddish brown. The crack between them is the darkest part.

Skin colour can be difficult to mix. Avoid making it too pink. Yellow ochre, white and Indian red mixed together make a good basic colour, which you can then adjust by adding small amounts of other colours. Add green or blue to take away the pinkness.

Skin reflects the colours around it.

A child's head is usually rounder than an adult's. The jaw is softer and less pronounced.

Lighting

Side or three-quarter lighting is usually best for a full face view.

A face looks more interesting when there are shadows to emphasize its features. Try to experiment with the position and strength of the lighting until you find the best effect.

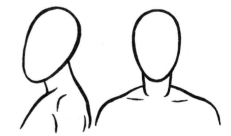

To do a portrait you need to draw in the main features first. When you are painting someone's head, do not be put off by the idea of trying to get a likeness but

concentrate on seeing the head as areas of colour. If you get the right colours in the right places the likeness will emerge.

The head is roughly egg-shaped. Notice the angle at which it fits on to the neck and the angle of the neck to the shoulders.

25

Outdoor scenes

Painting out-of-doors can be difficult to adjust to. There are usually a lot of distractions and the view can seem immense and complicated. Before beginning a painting it is a good idea to make a quick sketch to outline the important shapes and help plan the composition.

Water-colours are ideal for sketching and painting outside. Oil paints and acrylics can be used, but you need to be well organized and to allow a lot of time. It is better to use these at home, using sketches done outside to help you.

The countryside

Sweeping, panoramic views are often the most spectacular and inspiring, but they are also the most difficult to paint. If you do attempt a distant view, use a viewfinder* to help you compose your picture. It helps to give a sense of distance if you find a road, fence or stream to lead your eye into the picture.

Trees

When painting trees, do not try to put in too much detail. In winter, when you can see the branches clearly, it is better to paint the trunk and main branches quite carefully, rather than roughly dashing in all the small branches and twigs, which appear as solid colour from a distance.

Trees in summer have more definite outlines. Their shapes are softened by the foliage. Try and see them as areas of light and dark colour. Notice that there are often patches of sky showing through the leaves.

The shapes and patterns made by close-up views of things like this wall can make an interesting painting. Keep a look out for good subjects in unexpected places.

The colours and shapes of ploughed fields and crops can create a beautiful patchwork effect, which often works well in a painting.

Colours in landscape

You need to mix your colours very carefully when doing a landscape painting. Unmixed colours very rarely resemble the natural colours of the countryside. To help you work more quickly you might find it useful to mix a range of greens, browns, greys and blues to start with and make the necessary adjustments to them as you go along.

Remember that the colours you see will change according to the strength of the light. It is important to keep your painting looking fresh, so you have to decide on the light you like, try to remember it, and not keep changing the colours as the light changes.

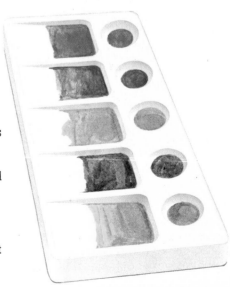

For more about viewfinders see page 17

Sky and clouds

Sky and clouds are a very important part of most outdoor scenes. The light and colour in the rest of the painting will be determined by what is happening in the sky. It can change very quickly so it is probably better to concentrate on getting the colours and the general shapes right, rather than the exact details of each cloud. Some types of cloud catch the light and have shadows just like any other object.

Water

Water reflects the colours of the sky and can change colour very quickly if the sky changes colour. It also reflects objects close to its edge. In strong light reflections can be almost as bright as the things they reflect. Painting reflections takes a bit of practice, especially if you want to paint ripples or running water. Treat them as coloured shapes, which get smaller as they get further away.

Towns and cities

Even the ugliest towns can be full of exciting views to paint. Street scenes provide endless possibilities. Reflections in such things as shop windows, cars and wet pavements can add extra interest to your picture.

If you have a window that looks out over rooftops you may find that the view will make a good picture. Chimney pots, aerials and wires can make very interesting shapes.

Parks are good places to paint in if you want to paint well organized landscapes with people in them. People move constantly, so if you want to include them you have to decide where you will put them in the picture and stick to that decision. If you want to paint crowds of people, markets and stations are good places, provided that you can find a quiet corner for yourself.

Industrial scenes can make very dramatic pictures. The colours are usually rather sombre and you have to look very carefully to distinguish all the different shades that are there.

Painting from imagination

Painting from imagination means painting pictures straight out of your head, instead of painting from life.

You may get ideas for paintings from things you remember, from your dreams or by combining fragments of things you have seen and heard about, or simply by thinking up ideas as you go along. The pictures you do in this way may be realistic or unrealistic.

It is very difficult to get exactly the effects you want from memory. You will probably find that you need to use real-life examples, pictures and photographs of certain things.

Realistic scenes

To paint a convincing realistic scene from imagination you need to make sure that your highlights and shadows are true to life, that the perspective and proportions are accurate, and that you choose your colours carefully.

Flat scenes

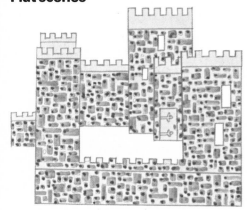

You can make very interesting unrealistic pictures with decorative shapes by ignoring perspective and painting things as though they were flat rather than solid.

Decorative paintings

Certain things, such as animals with distinctive markings, can make very decorative pictures, if you emphasise the patterns made by the markings.

Unrealistic colours

Painting things in totally unrealistic colours can create surprising and interesting pictures, which make you look at objects in a new way.

Distorted sizes

You can alter the sizes of familar, everyday, objects to create a weird and dream-like effect.

Unusual combinations

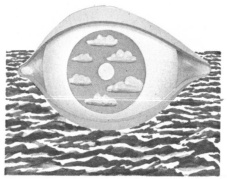

Unusual combinations of objects and scenery can also produce a strange, sometimes disturbing, effect.

Collecting useful pictures

You may find it useful to build up a collection of pictures on subjects you might want to paint. Cut them out of old newspapers and magazines and keep them in subject groups.

Abstract painting

There are two ways of setting about producing an abstract painting. The first is to choose an object and to concentrate on one aspect of it, such as the colour, or the shape, so that it no longer looks realistic. The second is to build up a painting from lines and shapes and blocks of colour that do not represent anything.

It can be great fun to produce an abstract painting. The ideas on the lower half of the page can help to get you started on one.

Working towards an abstract

Try choosing something like a tree or plant. First paint it very carefully just as you see it. Then paint it again, using what you feel to be the most important features. Finally, paint only the colours and shapes essential to show its character. You may want to use more than three stages.

Ways of getting ideas

Try doodling on a piece of paper, perhaps with your eyes closed, and see if any interesting shapes emerge. Use any shapes that interest you as the basis for your painting.

Drop a blob of ink or diluted paint on to a piece of paper. Fold the paper in half and press it flat, then open it out and see what shape the blot has made. You can add details to it.

Cut out different shapes of coloured paper and arrange them on a piece of white paper until you arrive at a design you like. Use this design as the basis for your painting.

Colours can create a certain mood or feeling. Choose a word such as "sadness" or "fear" and see if you can illustrate it simply by using colours and shapes. The picture above illustrates anger.

Put a piece of paper over a textured surface, such as the bark of a tree, and rub it with a pencil. The pattern and shapes that emerge may give you ideas to use in a painting.

Make up a set of rules for yourself. You might decide that you can only use straight lines, as in the picture above. See how many varied effects you can get without breaking the rules.

Frames and mounts

When you have done a painting you are pleased with you may want to complete it by framing it.

Picture frames come in a great variety of styles and can be made of aluminium, plastic or wood. You can have your pictures framed by a professional picture framer, or buy ready-made frames, or you can make them yourself. Once you have bought some basic equipment, it is far cheaper to make them yourself.

Paintings done on canvas can be hung without a frame. Paintings done on paper need something to stiffen them. Oil paintings and acrylics do not need to be covered by glass, but water-colours and gouache paintings usually have a piece of glass and a cardboard mount to protect them. The wooden frame described on this page would be suitable for use with or without glass.

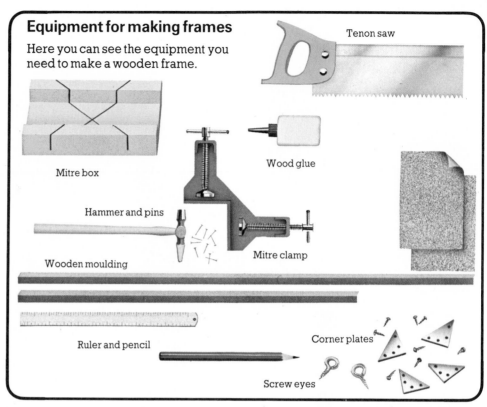

Equipment for making frames

Here you can see the equipment you need to make a wooden frame.

Tenon saw

Mitre box

Wood glue

Hammer and pins

Wooden moulding

Mitre clamp

Ruler and pencil

Corner plates

Screw eyes

1

Outside edge

When you measure the wood remember that the outside edge of the frame must be longer than the sides of your picture.

2

Use a mitre box when you cut the wood, to make sure you cut it at an angle of exactly 45° to the sides.

3

When making a corner, glue both bits of wood and wait until the glue is tacky before pressing them together.

4

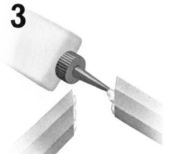

While the glue is drying, put the corner into a clamp to hold it in the right position.

5

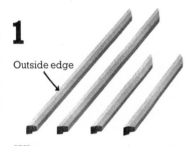

To strengthen the corner use pins. Tap them in carefully to avoid damaging the wood.

6

Do diagonally opposite corners first and then fit your two L shaped pieces together.

7

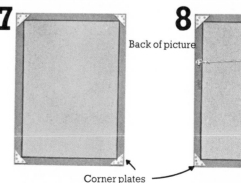

Corner plates

When you have fitted the picture into your frame, screw in corner plates to keep it in position.

8

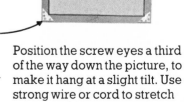

Back of picture

Position the screw eyes a third of the way down the picture, to make it hang at a slight tilt. Use strong wire or cord to stretch between the screw eyes.

Water-colour frames

Using a cardboard mount to frame a water-colour not only protects the painting but also sets it off to its best advantage. You can buy coloured mounting board at most art shops. Take your picture to the shop to make sure you choose a colour that suits it.

The best kind of mount is a window mount. To make one, use a metal ruler and sharp knife to cut a piece of mounting board the right size to fit the frame. Cut a hole in the centre the same shape as your picture but slightly smaller, so the edges of the paper are covered by the mount.

You can do without a frame and use clips to hold the glass and backing boards together. This is very effective for some styles of painting.

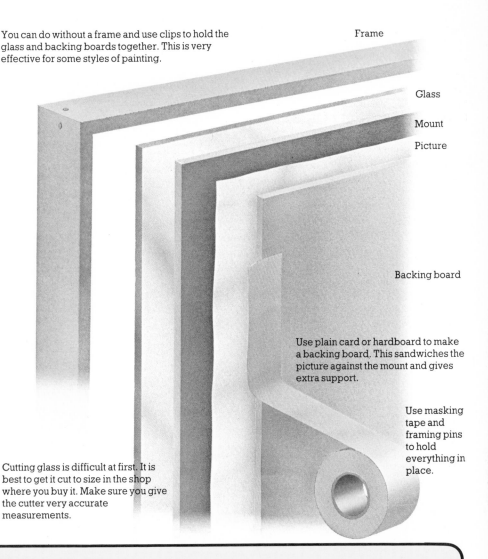

Frame

Glass

Mount

Picture

Backing board

Use plain card or hardboard to make a backing board. This sandwiches the picture against the mount and gives extra support.

Use masking tape and framing pins to hold everything in place.

Cutting glass is difficult at first. It is best to get it cut to size in the shop where you buy it. Make sure you give the cutter very accurate measurements.

Going further

When you are learning to paint it is very helpful to spend time looking at paintings done by other people and to think about how they have achieved their results. Looking at an original painting is very different from looking at a reproduction of the same painting. Most cities have at least one art gallery or museum with a collection of pictures. If you do not live near a museum or gallery looking at postcards and pictures in books can also teach you a great deal. The *World of Art Library* (Thames and Hudson) is an illustrated series of books on individual artists and periods of art history. The *Colour Plate* series (Phaidon) has good large scale reproductions of famous paintings.

Most libraries and bookshops have a wide selection of books aimed at helping you to learn how to paint. Many suppliers of artists' materials also stock a small selection of practical painting books.

The following books may help to take you beyond the first basic stages of learning described in this book:
Painting by James Ogilvie-Forbes (Ward Lock).
Oils and *Watercolours*, two books by John Holden (Marshall Cavendish).
Introducing Oil Painting and *Introducing Watercolour Painting*, two books by Michael Pope (Batsford).

The *Painting and Drawing* series (Studio Vista) includes books on how

to use different types of paint and how to paint a wide range of different subject matter. Also in this series are *Colour for the Artist* by Hans Schwarz and *Composing your Paintings* by Bernard Dunstan.
The *Leisure Arts* series (Search Press) and the *Artists' Painting Library* (Pitman) both cover a wide range of more specialized subjects.

The following books are very comprehensive and detailed and quite expensive, but are useful reference books and worth looking for in a library.
The Artists' Manual (Macdonald).
The Complete Guide to Painting and Drawing Techniques and Materials (Phaidon).

Glossary of painting techniques

Here is a list of words which you may come across in other books about painting. Each of them describes a particular method of applying paint.

Alla prima: method of painting in oils without doing any underpainting. This is the normal method of oil painting these days.

Collage: using paper or other materials glued to a flat surface to create a picture.

Flat colour: technique of using opaque, smooth areas of colour, which do not show any brushmarks.

Frottage: oil painting technique, which involves using a piece of either flat or crumpled paper pressed into an area of thick paint and lifted off to create a mottled effect.

Glazing: acrylic or oil painting technique of using layers of thin, transparent paint placed on top of each other, so that the first colour shines through the others.

Hard-edge: acrylic painting technique. Using a mask, usually of masking tape, in order to paint absolutely straight lines.

Impasto: using a thick layer of oil or acrylic paint, that stands out from the surface of the painting.

Line and wash: water-colour technique. Line is applied with a fine brush or pen, using waterproof ink or paint and a wash is applied on top.

Masking: using a mask to cover areas of a painting, in order to create a particular effect. A mask can be made from special masking fluid, film or tape, board or plastic. With a water-based paint any water-resistant material, such as oil pastels or wax crayons, can be used.

Overpainting: gouache technique of painting opaque colour over an area of solid colour or over a colour wash.

Scumbling: applying paint with a nearly dry brush using a scrubbing motion. This technique is used for blending hard outlines or for breaking up a layer of colour so that another layer underneath shows through.

Staining: using a thin consistency of acrylics or oils to cover canvas with a layer of colour.

Stippling: using dots or blobs of colour, instead of brushstrokes, lines or flat colour.

Underpainting: using very thin paint to lay in the basic shapes and areas when starting on an oil painting.

Wash: a thin stain of water-based paint brushed over paper. A wash is normally used as a background to a painting.

Index

The

GEM
KINGDOM

by Paul E. Desautels

SPECIAL PHOTOGRAPHY BY

Lee Boltin

A Ridge Press Book
Random House,
New York

Editor-in-Chief: Jerry Mason
Editor: Adolph Suehsdorf
Art Director: Albert Squillace
Project Art Director: Harry Brocke
Associate Editor: Moira Duggan
Associate Editor: Barbara Hoffbeck
Art Assistant: Mark Liebergall
Art Production: Doris Mullane

Contents

Introduction

An incredible number of books has been produced through the centuries to enlighten, instruct, inform, entertain, and confuse people about gems. There is a fascination about gems that cannot be denied. As long as it exists, there will be books which attempt to feed it.

Generally, authors are serious in their effort to sort out the facts and stories about gems, gemstones, gemology, gem cutting, gem mounting, gem identification, gem history, and gem markets. Consequently, most books tend to restrict their topics to specialized subjects, since it seems impossible to put everything in a single work. Thus there are books on diamonds, on jade, on gem cutting, on jewelry, on opal, on amber, on gem lore, and even some which are encyclopedic. These, for example, list with abbreviated commentary the pertinent facts about the gemstones and their origins. Even so, no book ever tells the entire story and each new one is awaited with interest for what it may add.

This book is not intended to be exhaustive about gems and gemstones under any particular category. Rather, it attempts to probe into most aspects of the gem world. It illuminates the antiquity of man's interest in gems and touches on the interplay that gems have had with his affairs through all his history. Then, too, it spends some time on the gems themselves, where and how they are found, how they are identified and distinguished from one another, and how they are put to use. Because of the high intrinsic value of gems, fakes and substitutes have naturally played a part in the story. Their impact is ever-changing, as modern laboratory technology finds new and improved methods of imitating nature. New materials and tools and more widely disseminated information about gem-cutting and jewelry-making methods have encouraged experimentation leading to considerable improvement in the processes and their glittering products. Fashion and fancy change constantly, putting gem and jewelry cutters, mounters and merchants under considerable pressure to bend to the public will. The great collections of crown jewels, collections of antique jewelry, and certain of the more renowned gems and their mountings are constant reminders of the continuity of these ancient and respected crafts.

All of these topics are touched upon here to some extent. Gem lore (Chapter 1) is seen here as an expression of man's use of these pretty stones for jewelry, talismans, barter, medicine, investment, and for conspicuous display of wealth. The facts about gems are outlined (Chapter 2) in a discussion of their general characteristics as they are determined and organized by the science of gemology. Because gemstones are minerals they behave in every way like their fellow mineral species. And yet the science of gemology, closely allied as it is to the science of mineralogy, has accumulated its own font of knowledge and some of its own special techniques. Man-made gems, imitation and altered gems are given some attention (Chapter 3), along with the completely natural gemstones (Chapters 4 and 5), and some of the interesting sources from which they have come (Chapter 6).

There are differences in attitudes, of course, toward those gems considered popular, fashionable, and marketable, as contrasted with those only a connoisseur-collector could love. Accordingly, an admittedly arbitrary division is made of them here. A review of gem-cutting and polishing methods and styles to produce cabochons, faceted stones, and carvings is also given (Chapter 7). The volume of present-day literature directed toward thousands of gem-cutting hobbyists, outlining cutting technique and technology, reflects the growing popularity of what was at one time a professional craft. Since jewelry would be unthinkable without the noble metals—gold, silver, platinum, and their alloys—these are described in a section on jewelry (Chapter 8), which also includes some discussion of style. Some gems and jewelry have survived the ravages of time. The best-known of these are preserved in private collections and in some still-existing great collections of crown jewels which are here given a section of their own (Chapter 9).

To sum up the story of gems, jade (Chapter 10) is used as an example of a durable gemstone which has been man's companion for perhaps more than seven thousand years. It has been very important to and intimately associated with at least two great cultures and has brought out the best in man's gem-cutting abilities and aesthetic urges. Altogether, its story seems to embody most of the ideas basic to any survey of man's long fascination with gems.

Wherever this love affair between man and gems may lead, the hope is that this book will be a helpful guide. The illustrations have been carefully chosen and devised to add as much of a visual dimension as possible to the survey. For those who know little or much about gems it should provide some new insight into an old but always new story.

The author and photographer wish to thank Dr. George Switzer for an expert reading of the manuscript (while, of course, relieving him of responsibility for any errors that may remain), Dr. Joel Arem for photographic assistance and advice, and Ken Kay for his inventiveness in the solution of technical photographic problems.

Paul E. Desautels

Prized Through the Ages|1

Imagine the astonishment of the Spanish invaders of Mexico when they penetrated the royal palace of Tezcuco and saw its treasures. As reported in Prescott's *Conquest of Mexico,* they found "a courtyard, on the opposite sides of which were two halls of justice. In the principal one, called the 'tribunal of God' was a throne of pure gold inlaid with turquoises and other precious stones. On a stool, in front, was placed a human skull, crowned with an immense emerald, of a pyramidal form, and surmounted with an aigrette of brilliant plumes and precious stones. . . . The walls were hung with tapestry . . . festooned with gold rings [and] above the throne was a canopy of variegated plumage from the center of which shot forth resplendent rays of gold and jewels." Through all of man's known history, rulers, persons of great wealth, and those in positions of religious or political power have devoted much effort and money toward accumulating gems and objects made of precious metals. The Spaniards would have found similar treasures in almost any royal palace in the world.

Of course, the use of gem materials as a sign of wealth and position was old long before the conquest of Mexico. The importation of lapis lazuli from Turkestan into Egypt was a thriving business nearly a thousand years before the Hebrew exodus. Lapis lazuli and emerald ornaments found in Egyptian tombs provide the earliest definite evidence of extensive commercial gem traffic. The intense, violet-blue lapis lazuli, frequently speckled with golden bits of the mineral pyrite, was quite certainly the material known to the ancients as sapphire. Egyptian tombs, such as those in the pyramids of Dashour, have yielded objects designed in gold and lapis lazuli dating back twenty-four centuries B.C., to the time of the Twelfth Dynasty. All of it came from Badakshan, which was in ancient Media, in the Oxus valley. Very likely the Badakshan mines, now in Afghanistan, are the oldest operating mines in the world, having produced continuously for seven thousand years. By archeological evidence, the existence of the gem trade routes has been pushed back in time even to a predynastic period—well before 5000 B.C.

The Egyptian use of turquoise is of similar antiquity. During the First Dynasty (5300 B.C.), turquoise was being mined by Egyptians in the Sinai Peninsula, near Serabit-el-Khadem. Some of it made its way into four bracelets found in the tomb of King Zer dating from this time. Later Egyptian records give detailed descriptions of expeditions sent out to work the turquoise deposits of the Sinai Peninsula. Perhaps the greatest mining operation was that sponsored by Ramessu IV at Hammamat, in which eight thousand men were involved. Transportation of supplies to the sites was a major undertaking. One expedition required five hundred asses carrying food and water continually over a five-day round-trip route.

Seven or eight thousand years ago, when these activities took place, may seem like ancient times in the story of gemstone mining and usage. It is by no means the time of the beginning of man's interest in gems. Archeological evidences of his interest have been traced back until the trail is lost in the dim reaches of time.

The earliest available evidences of a sort of gem working come from Spain and southwest France. Here, the Ice Age artists of the middle Aurignacian period—seventy-five thousand years ago—were busily carving objects in bone, horn, ivory, and stone. Their most popular forms, as the art developed, were the cave bear, reindeer, mammoth, and other wild animals, but the earliest productions were obviously human figures. Before that time, man or man-like creatures may have accumulated and worked with special bits of stone to which were assigned some higher value.

The entire history, lore, and technology of gems is based on the fact that there are certain bits and pieces of natural material which, for one reason or another, are sought out and highly treasured by man for artistic and ornamental purposes. By and large these materials happen to be mineral matter. However, coral, bone, amber, pearl and other derivatives of living organisms have always been among them. In an Ice Age society pieces of workable bone or ivory likely had the same significance that mother-of-pearl has today. True, we normally tend to think of gemstones as superior varieties of minerals, but after thousands of years—it seems almost through special affection—the nonminerals ivory, coral, amber, jet, and pearl are accepted as part of the earth's treasury of gemstones.

Since the time of the Ice Age artists, man has been continuously and increasingly interested in the acquisition of gemstones and in the arts and crafts of their alteration to satisfy his conceptions of beauty and utility. Through the centuries gems have been the gifts and ran-soms of kings, the cause of many crimes of violence, of assassinations and thefts, and the overthrow of rulers. The stories of some particular individual gems span longer periods of history than do several of the great nations of the world. Gems have been used for many purposes, including personal adornment, investment, symbols of power, and gifts given out of sentiment. Their names and characteristics have even been used in literature to describe the delights of heaven. They have been used for jewelry, religious symbols, talismans, for barter, medicine, financial security, and as one of the simplest means of displaying wealth. Naturally, the legends and true stories about gems have accumulated in vast amount. A fascinating literature of gem stories exists, drawn from every aspect of man's existence from the time of his beginnings.

Favorite Ancient Gems

The availability and widespread love of turquoise was hardly limited to the Egyptians. Other peoples in other times made extensive decorative use of this soft, blue-to-green gemstone. The great treasures unearthed from the tombs of Monte Alban in Oaxaca, Mexico, included turquoise ornaments. Dating from about A.D. 1400, this trove contained a human skull encrusted with a mosaic of small turquoise plates. One great necklace has three strands of gold beads, three of pearls, three of coral, and fourteen of turquoise strung from two gold bars linked by gold bands. Turquoise inlay between carved figures was used on magnificent ceremonial knives made of jaguar bone. In addition to tomb remains, written

13

Opposite: Gold-sheet-covered decoration for head of Sumerian harp.
Left: Shawabty figure of Egyptian King Tutankhamen (c. 1350 B.C.) done in gold leaf with inlaid gemstones.
Below: Gold bracelet of same period with symbolic and decorative inlays of gemstones.

history records the value placed on turquoise by the subjects of Montezuma, the great Aztec ruler. He sent to Hernando Cortez, whom he first regarded as a long-awaited god, gifts of turquoise, a stone normally reserved for sacred use. These gifts included a mask, scepter, and earrings, all worked in turquoise mosaic.

Although turquoise had special appeal to early cultures, both undeveloped and advanced, it had rivals in amber and jade. Amber, soft and beautiful resin fossils formed from the secretions of living plants, has been universally popular since the time of ancient Greece. Early barrows, or burial mounds, excavated near Stonehenge at Salisbury, England, yielded amber plates strung as graduated beads that date from 2000 B.C. Sea trade from the Atlantic Ocean may have helped bring amber to the Mediterranean region. At any rate, by land or by sea, before the end of the Stone Age it had reached Bohemia and in the Bronze Age was traded all across central Europe to the Adriatic Sea. By 1500 B.C., ornaments made out of amber from the Baltic Sea were being included in the royal tombs of Mycenae in Greece. Some six hundred years later the Greek poet, Homer, gave the tawny substance its first mention as jewelry in literature. In the *Odyssey* he tells of a "necklace of amber beads" brought by a suitor of Penelope. By the time the Romans had begun to adorn themselves with amber, its possession had become a craze. One Roman writer of the first century A.D. indignantly reported that a small human figure carved in amber cost more than a good healthy slave.

Prepared beads and rough amber artifacts have been excavated in the Caucasus and in the Crimea. They seem to be of Baltic Sea amber, but since they were found with iron objects they are not so old as the Stonehenge amber platelets. Still, they date from about the time of Christ. There is a theory that the Phoenicians, great sea traders and navigators that they were, went on to the Baltic Sea for amber after sailing for tin from the mines at Cornwall, England. Baltic amber beads were often included in Tyrian tombs for several centuries before and after the time of Christ. Undoubtedly the distribution of the precious material throughout the world was due in some measure to Phoenician commerce.

Among the Gauls of ancient Europe, amber shared popularity with coral. These tribes had fairly easy access to this product of the sea because excellent coral was being harvested near Hyères on the Mediterranean coast of France, as well as near the Lipari Islands and off the coast of Sicily. Theophrastus, the Greek writer and student of Aristotle, about 300 B.C. described the nature of coral perfectly in his treatise on gems as "like that of stone. Its color is red, and its shape cylindrical in some sort resembling a root. It grows in the sea." Later authors, even on into the eighteenth century, were often confused about the animal nature and origin of coral.

Coral is not the only product of the sea valued in ancient times for ornamentation. Along with the gold and other gem-laden ornaments found at Monte Alban in Mexico was a number of large pearls. The largest was about the size of a marble, pendant-shaped, and with its luster fairly well preserved after five hundred years of burial. It must have been one of

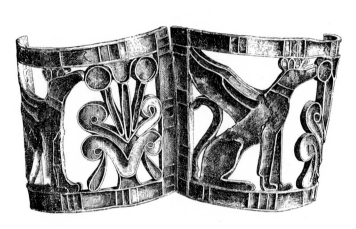

the world's largest and finest pearls.

Many people are aware of the often-repeated story of Cleopatra's use of the pearl to impress Mark Antony in the first century B.C. She is supposed to have tossed an extremely valuable pearl into her wine, swallowing the drink when the pearl dissolved. More than likely she swallowed it whole. Any liquid able to dissolve a pearl rapidly would have made short work of her insides. This costly pearl-swallowing act was not limited to Cleopatra. In the early part of the first century A.D., the demented Roman emperor, Caligula, was reputed to be a pearl swallower. These stories reflect the great value placed on pearls by the Romans from the beginning of the Christian era. The luxury of their acquisition became a disease, with prices rising to the equivalent of $300,000, reputedly paid by Julius Caesar for a single pearl.

Turquoise, amber, coral, pearl, and lapis lazuli are soft, easily worked gem substances. Hard stones, too, were valued in early cultures. Garnets and emeralds, satisfactory as hard, durable gemstones even by modern standards, were widely spread by early commerce. Most of the garnets came from the East. Trade with India very likely accounts for garnets found at Russian excavations in the Caucasus. The source was probably the same for garnets in ornaments found in excavations of Roman, Old English, Celtic, Etruscan, and Byzantine remains. A grave excavated at Sutton Hoo, England, thought to be that of Redwald, king of East Anglia about A.D. 600, was rich in garnet-trimmed ornaments. Among them is a sword with garnets encrusting its gold hilt. Some of the oldest Egyptian jewelry contained garnets. It is even suspected that the fourth gem of twelve in the breastplate of Aaron, the high

17

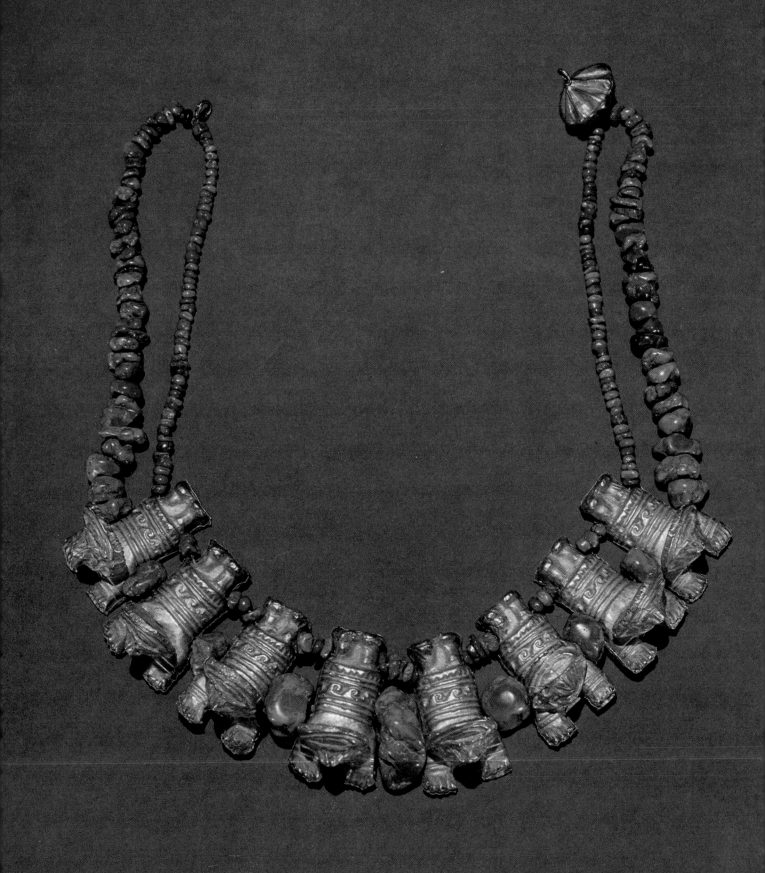

priest of Jerusalem (500 B.C.), was "nophek," or garnet.

The emerald is as ancient a recorded gem as any. In the oldest book in the world—the Papyrus Prisse in the National Library in Paris—appears the statement: "But good words are more difficult to find than the emerald, for it is by slaves that it is discovered among the rocks. . . ." The book was written about forty-five hundred years ago, but the statement was copied into it from a writing a thousand years older. The papyrus is referring to emerald that came from mines similar to those at Jebel Zabarah in Upper Egypt. Plentiful evidence of ancient mining activity still persists there. The hills are riddled with excavations and tunnels, while remains of stone temples and houses are abundantly supplied with objects from about 1650 B.C. Emerald mining apparently continued at this site for thousands of years. Strangely, the ancient emerald mines had been abandoned and forgotten for hundreds of years before they were rediscovered in the early 1800's. The poor quality of the few emeralds recovered at that time explained the practical reason for the death of the mines.

Spanish conquistadores first learned about emeralds from Colombia in the early 1500's, when the natives presented some as gifts. New World emerald mining, however, was an ancient occupation by that time. A prehistoric grave, excavated at Mate Esmeraldas in Ecuador, yielded a worked emerald. It was a fine, typical, six-sided crystal of rich green color which had been rounded at each end and marked by two drill holes to make a very attractive gem.

Magical and Mystical Gems

A story told by Dr. George Kunz, the famous American gemologist of the early 1900's, concerns words traded by Emperor Charles V and his jester. "What is the property of the turquoise?" asked the monarch. "Why," replied the jester, "if you should happen to fall from a high tower whilst you were wearing a turquoise on your finger, the turquoise would remain unbroken." This rather refreshing remark showed a disbelief in the magic powers of gemstones at a time when people generally attributed supernatural virtues to them. Closer to our time, an Apache medicine man would have been more in awe of the turquoise. In addition to being a sort of badge of office, it had various magical influences, such as assuring that an arrow or bullet would speed straight to its target.

Obviously, belief in the effect of gemstones on the affairs of man was not limited to any age or culture, but seems to have been part of every human society. We see this reflected in the ancient practice of giving gem materials religious or spiritual significance. The signs of the Zodiac, for example, were associated in ancient times with gems which supposedly assisted them in exerting their influence on mortals. Early Christians and Jews made use of the same idea in relating certain gemstones to the Twelve Tribes of Israel and to the Twelve Apostles. In several cases these ancient gem names are almost impossible to identify correctly. Reasonable translations at least give us some idea of the gems in vogue just before and after the time of Christ.

Sources of Amber: Copy of oldest (1677)
known map of amber fishing and
mining area of Samland, on Baltic coast.
Bottom left: Picture accompanying
map shows amber fisherman. Right: Amber
seeps from tree in 15th-century woodcut.

SIGNS OF THE ZODIAC	
Garnet	Aquarius
Amethyst	Pisces
Bloodstone	Aries
Sapphire	Taurus
Agate	Gemini
Emerald	Cancer
Onyx	Leo
Carnelian	Virgo
Chrysolite	Libra
Aquamarine	Scorpio
Topaz	Sagittarius
Ruby	Capricorn

TWELVE APOSTLES	
Jasper	Peter
Sapphire	Andrew
Chalcedony	James
Emerald	John
Sardonyx	Philip
Sard	Bartholomew
Chrysolite	Matthew
Beryl	Thomas
Topaz	James the Less
Chrysoprase	Jude
Hyacinth	Simon
Amethyst	Judas

BIRTHSTONES	
January	Garnet
February	Amethyst
March	Aquamarine or bloodstone
April	Diamond
May	Emerald
June	Moonstone, pearl or alexandrite
July	Ruby
August	Peridot or sardonyx
September	Sapphire
October	Opal or tourmaline
November	Topaz or citrine
December	Turquoise or zircon

Early Jewish cabalists suggested that twelve stones, each engraved with an anagram of the name of God, had mystical power over the twelve angels: ruby for Malchediel, topaz for Asmodel, carbuncle for Ambriel, emerald for Muriel, sapphire for Herchel, diamond for Humatiel, jacinth for Zuriel, agate for Barbiel, amethyst for Adnachiel, beryl for Humiel, onyx for Gabriel, and jasper for Barchiel.

In addition, gemstones were related mystically to the twelve months of the year, twelve parts of the human body, twelve hierarchies of devils, and so forth. We still pay economic homage to the twelve birthstones.

Struck by the coincidence of all the twelves, St. Jerome, at the beginning of the fifth century A.D., drew attention to this relationship between the twelve gemstones in the breastplate of the Jewish high priest, the twelve months of the year, and the twelve signs of the Zodiac. Even at this time, however, the idea of the twelve birthstones had not established itself. The tendency still was to attribute certain powers to individual gemstones. One simply chose according to one's needs—perhaps amethyst to ward off intoxication, agate for becoming agreeable and persuasive, bloodstone as a guard against deception, coral for wisdom, or malachite as protection against malign enchantment.

The practice of relating a gemstone to the birth month of the wearer is relatively recent. Somehow it gained momentum among the Jews in Poland in the eighteenth century and was probably an outgrowth of active interest in the significance of the original twelve stones of the high priest's breastplate. With some

20

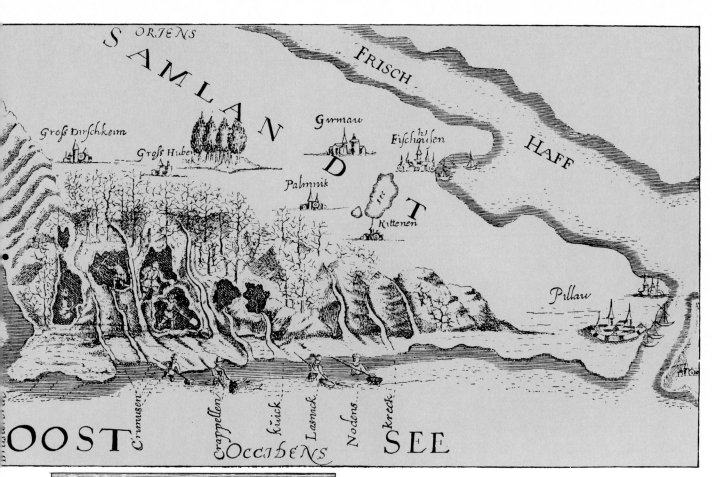

21

Gem Materials from Living Sources:
Polished Baltic amber with trapped insect,
19th-century drawing of branching coral,
American pearls from Long Island
Sound (A), New Jersey (B), Tennessee (D)
and Ohio (E) surround oyster shell.

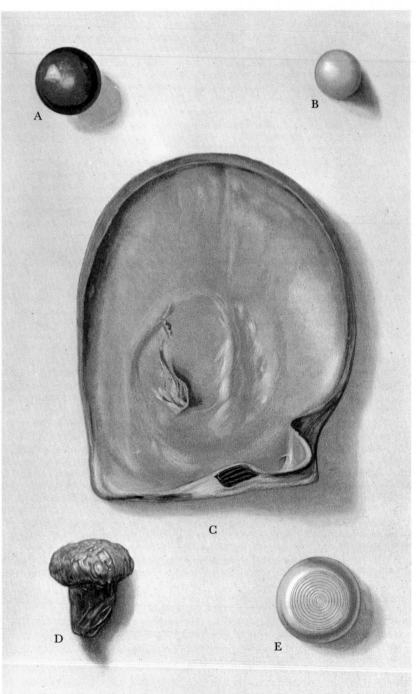

A

B

C

D

E

changes in the list, it became popular to choose a particular stone to represent each month. An individual could change ornaments each month, working through all twelve in a year and thereby receiving the virtues of all the stones. Gradually, emphasis changed to the wearing of one stone representing the birth month, thus bestowing an extra measure of its virtues on the bearer.

It wasn't always the gem material itself that was presumed to influence the affairs of men. Frequently, it was the carving or engraving which appeared on it. By medieval times the acceptance of the magic of engraved gems was complete. A fourteenth-century Italian manuscript advises that "a good stone is that one on which thou shalt find graven or figured a serpent with a raven on its tail. Whoever wears this stone will enjoy high station and be much honored; it also protects from the ill effects of the heat." This seems to be a very unusual, though highly practical, combination of influences.

George Kunz, in *The Curious Lore of Precious Stones,* notes other instances in a thirteenth-century book: "A frog, engraved on a beryl, will have the power to reconcile enemies and produce friendship where there was discord. A lion or an archer, on a jasper, gives help against poison and cures from fever. A bull, engraved on a prase [chrysoprase], is said to give aid against evil spells and to procure the favor of magistrates. The well-formed image of a lion, if engraved on a garnet, will protect and preserve honors and health, cures the wearer of all diseases, brings him honors, and guards him from all perils in travelling."

Medicinal and Therapeutic Gems

Unquestioning belief in the mystical characteristics of gemstones was coupled in early times with trust in their medicinal values. The name "jade," for example, was originally applied through a naïve belief in the stone's power. The Spanish conquerors in Mexico prized the jade pieces they got from the Indians, especially since they were considered a cure for kidney ailments. Because of this, they were called *piedras de yjada*—"stones of the side"—or kidney stones. From *yjada* comes our word jade. Similar notions about the curative and other powers of gems became common and still persist in less advanced cultures.

There were really two kinds of protective substances, those that functioned when taken internally or applied to parts of the body, and those whose mere presence warded off bodily ills or brought about cures. Without doubt, some of them were effective some of the time. The power of suggestion in curing certain psychosomatic disturbances was just as effective then as it is now. There is little difference in using a gem as the instrument to treat such an illness and in using a harmless and ineffective placebo pill, as is sometimes done today in standard medical practice.

Pliny the Elder (A.D. 23–79), in his great thirty-seven-volume *Historia Naturalis,* was highly skeptical of the curative powers assigned to the gems. Nevertheless, he faithfully recorded the ancient beliefs, which were quite in tune with the superstitions of his fellow Romans. (One wonders if that is the reason for the appearance of the same sorts of claims in

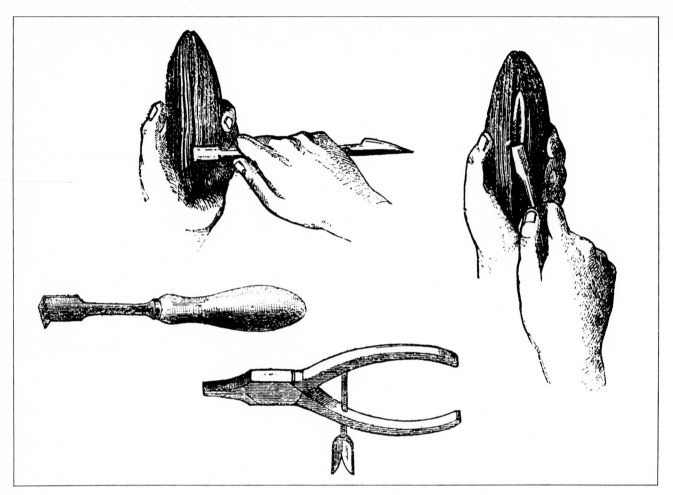

some present-day writing.) Totally aside from their chemical abilities, the ancients taught that red stones, such as carnelian, garnet, ruby, and spinel, were good for all sorts of bleeding and inflammation. The amethyst owes its very name to a Greek expression translated as "not to be intoxicated." Green stones were reserved for eye diseases. Emerald, however, was favored with additional powers; a sixteenth-century Spanish physician claimed to have cured dysentery with it. The sufferer had to wear one emerald touching the abdomen and carry another in his mouth. Obviously, the cure was limited to persons of wealth or those able to borrow a pair of emeralds from a friend.

Internal application of gemological remedies was another story. Among the more universally useful remedies, amber was effective internally and externally by influence or by actual application. Camillus Leonardus in his work, *Speculum Lapidum*, written in 1502, reports that amber "naturally restrains the flux of the belly; is an efficacious remedy for all disorders of the throat. It is good against poison. If laid on the breast of a wife when she is asleep, it makes her confess all her evil deeds. It fastens teeth that are loosened, and by the smoke of it poisonous insects are driven away." Leonardus and later authors gave amber all the powers of a modern antibiotic, with many additional virtues thrown in for good measure.

From the use of malachite as a powerful local anesthetic to sapphire for curing boils, there was some gemstone or combination of gemstones suitable for treating every ailment.

*Opposite: Tools for opening pearl
mollusks were urged by gemologist G. F. Kunz to avoid
destruction of shells and animals. Below: Henry
Philip Hope, of Hope Diamond fame,
also owned great 2-inch, 3-ounce Hope Pearl.
Right: 15th-century pearl merchant.*

Very likely there were also innumerable explanations for the frequent failures. Thomas Nicols in *Arcula Gemmea,* in 1653, warned practitioners of the art to "Beware of the use of gems (unless you be sure they are true) in physick, by reason they are so frequently adulterated."

Whatever the beliefs, man must have been comforted during these times to know that there were substances in the earth to which he could turn for help to influence his physical and spiritual environment.

The Breastplate and The Bible

Gemstone mythology may have taken form around the gem-studded breastplate worn by Aaron, first high priest of the temple at Jerusalem several centuries before Christ. Rabbinical writings of the time tended to include marvelous tales of gems associated with Abraham, Noah, and others of Biblical fame. Mohammedan religious legend, with its various heavens made of precious stones, added its contribution. Speculation and reverence for the breastplate, however, exerted very strong influence. The stones of the breastplate appear as one of three major lists of gems in the Bible. One of the other lists is given in Ezekiel 28:13 of the nine stones used by the King of Tyre to adorn his robe. The third list, in Revelations 21:19-20, gives the twelve foundation stones, each one carrying the name of an apostle upon which the new Jerusalem would be built.

One thing is fairly certain about the list in Exodus 28:17-20 of gems in the breastplate. The kinds of stones they were will probably

STONE NUMBER	HEBREW NAME	SEPTUAGINT (GREEK) CIRCA 250 B.C.	JOSEPHUS (GREEK) CIRCA A.D. 90	VULGATE (LATIN) CIRCA A.D. 400	AUTHORIZED VERSION A.D. 1611	PROBABLE STONE (BY MODERN NAME)	TRIBE
1	Odem	sardion	sardonyx	sardius	sardius	carnelian	Reuben
2	Pitdah	topazion	topazion	topazius	topaz	peridot	Simeon
3	Bareketh	smaragdos	smaragdos	smaragdus	carbuncle	emerald	Levi
4	Nophek	anthrax	anthrax	carbunculus	emerald	garnet	Judah
5	Sappir	sappheiros	iaspis	sapphirius	sapphire	lapis lazuli	Issachar
6	Yahalom	iaspis	sappheiros	jaspis	diamond	rock crystal	Zebulun
7	Leshem	ligurion	liguros	ligurius	ligure	zircon	Joseph
8	Shebo	achates	amethystos	achates	agate	agate	Benjamin
9	Ahlamah	amethystos	achates	amethystus	amethyst	amethyst	Dan
10	Tarshish	chrysolithos	chrysolithos	chrysolithus	beryl	citrine	Naphtali
11	Shoham	beryllion	onyx	onychinus	onyx	onyx	Gad
12	Yashpheh	onychion	beryllos	beryllus	jasper	jasper	Assher

THE BREAST PLATE of the JEWISH HIGH PRIEST.

3.   2.   1.

לוי   שמעון   ראובן

6.   זבלון   יששכר   יהודה   4.

9.   דן   בנימן   זם   7.

12.   אשר   גד   נפתלי   10.

11.

GEMS AND PRECIOUS STONES.

1. Carnelian. *East Indies.*
2. Peridot. *Levant.*
3. Emerald.
4. Ruby. *Burma.*
5. Lapis-lazuli. *Persia.*
6. Onyx. *East Indies.*
7. Sapphire. *Ceylon.*
8. Banded Agate. *East Indies.*
9. Amethyst.
10. Topaz. *Ceylon.*
11. Beryl. *Ceylon.*

12. Jasper. *Persia.*
13. Asteria. *(star-sapphire)*
    Ceylon.
14. Moonstone. *Kandy, Ceylon.*
15. Sunstone. *Norway.*
16. Alexandrite. *Kandy, Ceylon.*
17. Alexandrite. *Kandy, Ceylon.*
    (by artificial light).
18. Cats Eye. *Precious.*
    Chrysoberyl variety,
    Ceylon.

19. Precious Garnet. *(Pyrope)*
    Arizona.
20. Chrysoprase. *Silesia.*
21. Quartz. *(Rock Crystal)*
    Arkansas.
22. Bloodstone. *(Heliotrope)*
    East Indies.
23. Tourmaline. *Paris, Maine.*
24. Almandite Garnet. *India.*
25. Pearl. *(Oriental) Ceylon.*
26. Turquoise. *New Mexico.*

27. Demantoid. *(Green Garnet)*
    (Bobrowska Garnet, Uralian Emerald
    Ural Mountains, Russia.
28. Essonite. *(Cinnamon Garnet)*
    Ceylon.
29. Spanish Topaz. *(decolorized)*
    Spain.
30. Hope Blue Diamond.
31. Tiffany Yellow Diamond.
    (weight 125 ½ carats) S.
32. Fire Opal. *Mexico.*

never be accurately known. Only an unlikely recovery of the original breastplate could solve the riddle. The first breastplate, as described in Exodus, if it ever existed, was lost during the captivity of the Jews by the Babylonians. When the kingdom of the Jews was restored a new breastplate was prepared—perhaps about 500 B.C.—supposedly an exact copy of the original except that the Hebrew inscriptions on the gems were probably replaced by Chaldean or Syro-Chaldean. There is good reason to believe that the gems themselves were not duplicated. The sack of Jerusalem by Titus put the great temple treasures, including the breastplate, into the hands of the Romans. After a triumphal procession through the streets of Rome the treasures were placed in the Temple of Concord. There they sat until Rome fell to the Vandal Genseric in A.D. 455. From this point on, tracing the breastplate becomes a game of speculation. Some evidence suggests that the Vandals took it to Constantinople and that the Emperor Justinian later returned it to Jerusalem, where the Persians took it in the capture of the city in 615. It may be somewhere in Persia even now. However, the Arabs conquered the Sassanian Persian king twenty years later and may have made off with it. All this is speculation. The breastplate is gone but its existence is carefully—and confusingly—recorded. The confusion arises from several causes. For example, there were two breastplates—the Biblical original and the later duplication. The gemstones of the first could only have been those known among the Egyptians 1300 or 1400 years before Christ. By the time of the second, the gemstones reported by Theo-

phrastus and Pliny had appeared and more was known about cutting harder materials. The fact that the names used in the Bible are not the same names we use now for several of the gemstones compounds the confusion. The result is that diligent search through the Bible and other ancient writings, as well as study of artifacts, has produced differing interpretations of what gems were used. Each gem was inscribed with the name of one of the Twelve Tribes of Israel, but the proper order of names is not known for certain. Good but only indirect evidence of the order comes from other ancient writings. Perhaps the best sum of various interpretations and suppositions and voluminous research places the gems with their order, kinds, and engraved tribe names as shown on the table on page 26.

From the beginning, then, these gemstones were closely associated with many fanciful religious ideas. Josephus, the Jewish historian (A.D. 37-95), described the second breastplate as he actually saw it in use. His description, with its changes in the order of the stones, described with racial pride the power of the gems worn by the high priest to signal the presence of God at sacrifices or even to predict victory in battle. He was careful to point out the loss of this power two hundred years before his time because of God's displeasure with the Israelites. The Bishop of Constantia, almost four hundred years later, embroidered the statements of Josephus and attributed oracular power to the high priest's gem adornments. By changing color the stones could signal approaching death from disease or bloodshed as punishment for sinful behavior; or, by their resplendent purity, they could reflect good behavior. By the Middle Ages these ideas had developed into a mass of superstitions about planetary, medical, and alchemical influences.

Beginnings of Gemology

Gemstones, with certain minor exceptions, are only specially selected samples of mineral substances found in the earth. A full appreciation of their nature did not come until the development of the science of mineralogy. The original attractions of gems were their color, other aesthetic qualities, rarity, hardness, and susceptibility to a polish. Centuries had to pass before mineralogical distinctions could be made, before names could be applied consistently to various minerals, and before true exchange values could be established. During this period many sources of gems were developed and large numbers of gems flooded into the centers of culture to excite inquiry and speculation. There is much artifact evidence of a general, if moderate, knowledge of gems long before the early writers recorded it. This knowledge was diffused through the channels of commerce. Theophrastus, already mentioned as being knowledgeable about gems, is credited with the first attempt at classifying gem information. His work, and that of his master Aristotle, survived only in part. The writings of Pliny, especially the thirty-sixth and thirty-seventh volumes of his *Historia Naturalis*, comprise the earliest surviving gem record of major importance. It represents the accumulated wisdom of the time and gives us insight into the current condition of mineralogical science. In it, names are given

to mineral substances, distinctions are made among them, and general attitudes of criticism and observation are encouraged. The basis for classification of mineral knowledge was established for some time to come. Minerals were to be related according to the metals that could be extracted from them, the industrial or medicinal uses to which they could be put and their natural appearance or behavior. Unfortunately, the *Historia Naturalis* became a model for the metaphysical approach to science, a confusion of reality and myth which reached its extreme in the Middle Ages. Pliny's statements about amber are an excellent illustration of the problem. He firmly denied its source as the congealed tears of the Heliades, sisters of Phaeton who wept for the death of their brother and were turned into poplar trees by Jupiter. Just as firmly he denied the story, attributed to Demostratus, that amber, or "lyncurius," was the solidified urine of the lynx. Some writers feel that Pliny only compounded the problem by interpreting the lyncurius mentioned by Theophrastus as amber when it should have been recognized as what we now call zircon. Not satisfied with denying these myths, Pliny took a positive approach by flatly stating that "electrum," or amber, was congealed tree sap carried to sea by storms and later washed ashore. He was close to the truth. Yet, in spite of this note of understanding, he promptly ascribed to amber a collection of mystical powers enabling it to cure fevers, blindness, deafness, etc.

Time wore on and learning slowly advanced. The Arab scholar Avicenna (A.D. 980-1037) coincidentally wrote of medicinal uses

31

of minerals and helped to keep interest in these substances alive. Another Arab scholar, Averrhoes (A.D. 1120-1198), added little in his writing but served a similar function. Many other writers contributed to this state of affairs in which every tale was important, every gem had magical properties, and everyone was permitted to invent nonexistent gems for some special purpose. The gem notes written to Emperor Nero by Evax, king of Arabia, the gem accounts of the Bishop of Seville in the year A.D. 630, and of Bishop Marbodius of Rhennes as late as the sixteenth century were typical contributions to the lore of gems.

Critical speculation about mineral substances was finally undertaken by Albertus Magnus in his book, *De Mineralibus et Rebus Metallicis*, in the fourteenth century. He decided that "stones therefore have not spirits" but are endowed with "celestial virtues." This was still in keeping with the popular feeling about alchemy and metaphysical science in general. In 1556, Georg Bauer, better known by his pen name, Agricola, published the famous *De Re Metallica*. It brought up to date man's knowledge of mineralogy, mining, and the uses of metal ores. Agricola was the first earth scientist in modern times to apply critical judgment to the mineral kingdom. He adopted a system of classification under which minerals were earthy, saline, ignitable, or metalliferous. These categories were further subdivided by the varying characteristics of minerals in each group. Certainly this was the beginning of mineral systematics, which is now developed to a high degree. Like many of his predecessors, Agricola was fascinated by the regular forms

and interesting geometry of natural crystals.

After Agricola the pace of advancement accelerated. Hieronymus Cardanus wrote about the nature and forms of gems in 1663. He speculated on the causes for the geometric shapes of crystals and attributed them to some natural force implanted on the crystal, forcing it to grow in certain shapes. Already a glimmering of the truth was sensed. Caesalpinus, Gesner, and Kentman, at the end of the sixteenth and beginning of the seventeenth centuries, grew crystals in the laboratory and studied them. Many others began to work with minerals, gems, and crystals, thus beginning to unravel the true story. There were throwbacks: Peter von Arles, at the beginning of the seventeenth century, was still trying to relate the seven metals with seven precious stones and the seven planets, with the moon symbolizing silver and quartz.

In the middle 1600's, Nicolaus Steno, with his astute observations, gave crystallography its first meaningful series of measurements. The law describing angular relationships between natural crystal faces is still often referred to as Steno's Law.

Later, in 1774, the professor Abraham Gottlob Werner, of Freiberg, in Saxony, published his ideas about the separation and grouping of known minerals in an essay on the "External Characters of Minerals." Separating and arranging minerals systematically, according to their color, hardness, inflammability, and so on, had an appealing logic, and Werner's students, who came from all corners of Europe to study with him, helped disseminate his notions. Mineralogical relationships were easier to establish and new predictions about their chemical constituents and physical behavior became easier to make. In our time, the chemical and physical composition of minerals is the only evidence by which we distinguish between species.

By the middle of the eighteenth century, the allied sciences of mineralogy and chemistry were under intensive investigation all over Europe. The transition was being made from superstition, myth, and conjecture about minerals, to a search for and reliance on observed facts and verified information. The mental forces of critical inquiry and experimentation had taken hold.

Gems Today

Although gems have had various uses and meanings at different times in the course of history, they have been highly prized at all times. This is as true today as it ever was. Gems combine high value with small size. They are easy to transport, easy to conceal, and easily convertible to money. In times of stress, when the value of money was weakened, gems have often been the means of saving the value of an estate. Real-estate holdings may not survive a revolution, but a secret hoard of gems may outlast any political system. Gems also are used as an instrument of investment. With expanding populations and increasing prosperity in many countries of the world, the number of potential customers for gems—as for paintings and other works of art—is increasing far more rapidly than the available supply. This has the effect of steadily driving up prices.

Totally aside from their use in trading and investment, the greatest part of the vast annual sale of gems is for use as personal adornment or in works of art. Gems and the jewelry in which they are mounted remain desirable throughout the years, even though styles may change as cutters and designers contend with each other to create better and more beautiful pieces. The new popularity of colored stones in the United States has also helped to stimulate new energies in the gem industry, which in the past has been almost totally tied to diamonds.

Activity in the gem market has been accompanied by a thirst for knowledge about gems, yet the public is largely ignorant of the facts about gemstones. The tendency of buyers is to rely on a jeweler of established reputation and honesty, just as one would rely on an accredited physician or surgeon. Various groups have been founded for the purpose of insuring the respectability of the gem trade: the American Gem Society, the Gemological Institute of America, and others, either police the commercial activities of their members or offer them technical advice. All have done much to insure the public against fraud or poor advice. Nevertheless, there is a need for the dissemination of information about gems to the public to satisfy curiosity and to stimulate interest. Much of the knowledge needed is so technical that it hardly seems possible for the potential gem buyer to master it. Fortunately, many basic facts and ideas about gems and jewelry are easily assimilated even though the trained jeweler is still the authority in technical matters. No longer the treasures of the rich and powerful alone, gems and a true knowledge of them are now available to all of us.

The Qualities of Gemstones | 2

The science of gemology is concerned with investigating and establishing facts about gems and gemstones. Somehow, because gems usually are objects of monetary value, it is often difficult to think of them in scientific terms. Questions about where they come from, what they are made of, and how they can be distinguished from one another usually take second place to questions about their value. However, these other questions must be answered first, before decisions can be made about value. For example, the market values of look-alike quartz and topaz differ greatly. Proper identification of a suspected topaz is crucial before a relative price can be assigned to it. Aside from pricing problems, other kinds of serious mistakes can be made because of faulty identification. Two of the more famous rubies in the world, the Black Prince's Ruby and the Timur Ruby, are not rubies at all, but spinel. Even so, they are extremely important stones because of their historic past and their prominence among the British crown jewels.

The science of gemology is quite different from the lapidary art, which deals with techniques for cutting, polishing, and generally shaping gemstones for ornamental use by themselves or in jewelry. Cutters and carvers and craftsmen through the centuries have refined the techniques and equipment for gem cutting. Lapidary work has become a joint venture for artist and craftsman, and it leans heavily on the science of gemology. Certainly, a lapidary must know how his gemstone is going to behave and how to turn its characteristics to advantage before he begins to work on it.

To most people who talk about gemology and about gems in general it becomes obvious that, through careless usage, some of the basic vocabulary has become confused. To set the record straight, "gemstones" are the specially treasured minerals found in the earth and "gems" are the objects fashioned from them. "Jewels" are gems that have been prepared for mounting in jewelry or other objects of art.

Why are only certain natural mineral samples specially treasured as gemstones? Because gems can be cut from them that have at least some of the qualifying characteristics: brilliance, beauty, durability, rarity, and portability. If the gem also happens to be "fashionable" it acquires status. Partial qualification is more often the rule, because seldom does a gem have a large measure of all these attributes. A good diamond with its high degree of brilliance and fire, as well as extreme hardness and rarity, comes close to the ideal. Opal, being relatively fragile and having little but its rarity and breathtakingly beautiful play of colors to offer, still qualifies. A number of gemstone species, such as the beautiful blue Tanzanite, have produced beautiful cut gems, but their commercial success has been sharply limited by insufficient supply—or rarity carried to an extreme. With certain notable, fashionable exceptions, a gemstone can't afford to be too rare. Scarcity does not make a stone less important as potential gem material. It merely points up the strong effect on gem marketability of the accidental, uneven, natural distribution of these species in the earth. When supply is adequate, certain attractive gems, such as spinel,

Preceding pages: Becoming adept at
gem identification requires broad experience
with uncut gem materials (such as
large topaz crystal here), with cut stones,
with simple—but sometimes expensive—
tools, and with gem literature.

still do not compete as they should with other more plentiful kinds—tourmaline, for example—that exhibit similar ornamental characteristics. Still other mineral species in adequate supply, such as fluorite and sphalerite, which are beautiful when cut and are prized by collectors, are entirely too soft, are too easily broken or cleaved, or have some other physical weakness which makes them useless as commercial gemstones. Through a complex combination of accidents, then, certain mineral samples assume an intrinsic importance as gemstones. Continents have been explored, wars fought, crimes committed, fortunes made and lost, all in pursuit of these uncommon bits of minerals.

Since gemstones, with a few notable exceptions, are minerals found in the earth's crust, the laws and procedures applied to the study of minerals fit them perfectly. Any trained mineralogist can soon become a competent gemologist, since he is already familiar with the techniques of identification and knows the fundamental chemistry and physics of natural substances. A mineral is a natural substance having a definite chemical composition and definite physical characteristics by which it can be recognized and distinguished from other substances. Technically, in mineralogy, those natural substances formed by living organisms are excluded. This means that amber and jet, formed by plants, and coral and pearl, produced by animals, are not minerals. However, all four are traditionally included among the gemstones, because they qualify on grounds of beauty, rarity, etc.

The gem mineral's characteristics of brilliance, beauty, and durability arise directly from the kind of chemical composition and also from the kind of internal atomic structure it has. Sometimes, natural accidents of growth and the introduction of impurities during the formation of the gem minerals may enhance their interest and value. On the other hand, severe accidents of growth may destroy their usefulness. Obviously, some understanding of the chemical and physical reasons for mineral characteristics is needed to appreciate and understand gemstones.

Chemical Composition: We know that the universe is made up of a relatively few basic building materials, the hundred-odd chemical elements. Some of their names—gold, silver, copper, sulfur, and oxygen—are very familiar. Others such as beryllium, zirconium, and boron sound less familiar but are important among gem minerals. Still others are so rare as to be of no importance or interest in this discussion. A small number, perhaps twenty-five, supply materials to make up all significant gemstones. A few more, present in tiny trace amounts, may impart color or other occasional special characteristics.

The elements which go into making up a mineral exist as innumerable, extremely small bodies called atoms. Each kind of atom—e.g., silicon or oxygen—has its own characteristic size and its own particular ability to join with other atoms. In nature, under various temperatures and pressures and in different mixtures, the elements are brought together and combine with each other to form minerals. Since 46½ percent of the earth's crust is oxygen and 27½ percent is silicon, it is not surprising that most minerals contain these two elements.

37

Three gems from Smithsonian collection
show differences of color and composition.
Step-cut, 103-carat zircon (top) from
Indochina; round-cut, 46-carat spinel from
Ceylon (bottom left); and square-cut
69-carat spodumene from Brazil.

Aquamarine, emerald, tourmaline, topaz, zircon, peridot, spodumene, and garnet are gem minerals known by the general name "silicates" and contain both silicon and oxygen as their major constituents. Ruby, sapphire, chrysoberyl, and spinel are "oxides," containing oxygen as a major constituent. As already mentioned, each kind of atom is limited in the kind and number of other atoms it can join. The chemist, through the years, has learned to predict the possible combinations and has developed very accurate methods of checking them in the laboratory. He can make almost innumerable combinations of elements in the laboratory, predicting in each case how they will combine with each other. He can also determine the kinds and relative quantities of atoms present in a mineral sample and use a standard method of noting them. His typical analysis of a mineral might show that there are equal numbers of zirconium (Zr) and silicon (Si) atoms present and four times as many oxygen (O) atoms. His notation, then, would read $ZrSiO_4$. This is the chemical notation or formula for the gem mineral zircon. Thus, the formula for aquamarine is $Be_3Al_2Si_6O_{18}$—a beryllium aluminum silicate; for chrysoberyl it is $BeAl_2O_4$—beryllium aluminum oxide.

All this seems simple enough until it develops that the chemical formulas for ruby and sapphire are identical—Al_2O_3. If absolutely pure, this aluminum oxide, Al_2O_3, is colorless. Ruby, however, is red, and sapphire by definition is any color except red. As the formula indicates, they are actually the same mineral, but ruby is aluminum oxide containing very small traces of the element chromium which cause it to have the red color. Sapphire seems to get its colors from tiny traces of iron or titanium, or both together. Certainly, this is a case where chemical impurities gathered by a mineral during its formation produce highly desirable results.

Internal Structure: The particular combining abilities of each kind of atom go a long way toward determining what combinations or compounds are possible. At the time a mineral forms, there are restrictions relating to the size, characteristics, and numbers of atoms present. Atoms are energetic, and exhibit this as rapid, erratic motion. As they rush about at phenomenal speeds they tend to fasten onto each other by strong attractive forces. Many trillions of atoms may pack themselves together this way in the course of an hour during the formation of one of these mineral solids. This would suggest that they all end up in a great, unstable, chaotic mass. Instead, because of the uniform distribution of attractive forces and relatively uniform sizes, they line up in remarkably orderly, repetitious, geometric patterns and hold themselves quite tenaciously in these patterns called crystal lattices. A good demonstration of how this happens can be prepared by shaking up a basketful of tennis balls. They all quickly settle down into an orderly geometric stacking pattern as they come to rest against each other. Nature permits surprisingly few stacking patterns, and all solid mineral crystals prove to have their atoms arranged in one of fourteen basic patterns, or combinations of these patterns. In any such pattern a foreign atom or impurity atom would have to have nearly the same size and attractive power as the others in order to fit into the structure. Atoms too large or too small are rejected and cannot enter the combination. It is not unusual to see iron atoms substituting for manganese atoms in some structures and chromium substituting for aluminum in others. Each member of the pair is quite close to the other in size and attracting ability and is, therefore, not rejected by the structure.

Crystals: Since the internal organization of mineral solids is so orderly, there should be, and frequently is, some external evidence of this order. If the solid crystal grows without external interference it will end up with flat, shiny surfaces, or crystal faces, that are parallel to layers of atoms within. It is even possible, by measurement of the angles which these external faces make with each other, to learn something of the atomic arrangement within. This is the business of the crystallographer. Since gem material is usually of high purity and has normally formed under ideal conditions in the earth, it is often found in transparent, well-formed crystals with good external faces. Often they are so well formed with regular, shiny faces that they give the impression of having already been cut and polished as gems. In size they may vary from microscopic individuals up to single crystal monsters weighing several tons each. Good, clear, clean, gemstone crystals, except for quartz, seldom are found in large sizes.

All solids, with minor exceptions, have orderly internal atomic arrangements and so are classed as crystals. The modern crystallographer studies these arrangements by using specialized X-ray techniques. Also, less fre-

quently, he will study the morphology or external shape and symmetry of these crystals. The science has advanced to the point where it is possible to determine just which kinds of patterns and symmetries may occur in natural and man-made crystals. All crystals, it has been discovered, can conveniently and logically be divided into thirty-two different kinds of symmetry groups or crystal classes. For the sake of further convenience and simplification, it is possible to group these thirty-two classes into six crystal systems, all classes in a given system having some important symmetry in common. The systems are: isometric, tetragonal, orthorhombic, monoclinic, triclinic, and hexagonal. The gem minerals can often be identified and distinguished from each other by the crystal systems into which they fall as they grow according to the dictates of their atomic structures. Beryl, in all its varieties such as emerald and aquamarine, is hexagonal, as is corundum with its varieties ruby and sapphire. Spinel is isometric, like diamond and garnet, while topaz is orthorhombic and zircon is tetragonal.

Everything that a gemstone is, how it looks, how it wears, and how it takes cutting and polishing, depends directly on its chemical composition and its internal structure. What a gemstone's characteristics are and how they arise makes an interesting study.

Light Characteristics

Light is easier to understand in terms of what it does than of what it is. The things it does are about as varied as the things it touches. Each substance has its own unique composition and structure, and handles light in a different way. When the word light is used it usually means visible light. Visible light is actually only a very small part of the total energy which is of a type known as electromagnetic radiation. Solid substances have an effect on all of this electromagnetic radiation. Their manipulation of visible light is of greater interest when dealing with gemstones. Sometimes the reaction between gemstones and visible light produces phenomenal and often beautiful or exotic results. Some of these which will be discussed here are color, schiller, asterism and chatoyancy, fire, and fluorescence.

Color: Color, or perhaps even its absence, is often the most striking characteristic of a gemstone. Basically, the color of a solid object depends on how it absorbs, transmits, or reflects the various wavelengths or colors of light to which it is exposed. "White" light is composed of a whole series of wavelengths or colors. When it strikes a solid object some of the colors may be absorbed by the atomic structure as added energy. If this absorbed energy is converted to heat and then dissipated it is, in effect, lost to the observer. However, there is the possibility that it will be converted to a different visible wavelength and passed on again. Assuming that some of it is absorbed, some or all of the remaining wavelengths are rejected or reflected at the surface and come back to the observer as something other than white, because some wavelengths have been subtracted from the original. Perhaps some or all will travel through the gemstone instead of being reflected back. The observer will see this as a color depending on the mixture of wavelengths which comes through the stone to his eye. If all wavelengths are absorbed, no light is reflected or transmitted and the solid looks black. If all are partially absorbed in an equal amount, the stone reflects or transmits gray.

A mineral whose own particular composition and structure are responsible for the colors that are passed through or reflected back is said to be "idiochromatic." The gemstone peridot is idiochromatic because it is always green and must be green because of what it is, chemically and physically. On the other hand, beryl can be yellow, green, blue-green, colorless, or even pink. Absolutely pure beryl is colorless. The colors in this case arise from the presence of trace amounts of impurities. Minerals with variable color depending on trace impurities, or sometimes on structural imperfections, are said to be "allochromatic." Both idiochromatic and allochromatic colors are caused by the interaction between light and some of the electrons that are part of the atoms that compose the mineral or are present as impurities. Atoms whose electrons commonly cause color in solids are usually of a group of elements called transition elements. These include copper, iron, manganese, chromium, nickel, cobalt, vanadium, and titanium. When these coloring agents, or "chromophores," are present in a gem the color is quite stable, but it can be changed under any conditions, such as excessive heat or radioactivity, which will change the nature of the chromophore.

Interference: Some gemstones, such as moonstone, are "pseudochromatic." That is, they exhibit colors caused neither by what they are made of, nor by the arrangement of their atoms. The colors show because some accident of growth has caused the development of various kinds of films or layers. Interference colors are the usual result. In describing the process of interference, the usual procedure is to invoke an image of the rainbow play of colors on a thin oil slick on a rain-wet street. A ray of light strikes the thin layer of oil at some angle. Some of the ray is reflected immediately from the top surface of the oil; some penetrates the thin film, and, in turn, is reflected from the contact surface where the oil film rests on the water. This second ray portion, traveling back through the oil film, continues on its way parallel to the first ray fraction. However, it is retarded because it has traveled a slightly longer distance. This means that the light waves in the two parts bouncing back have gotten out of step with each other. Since light waves are additive, the resulting combination of out-of-step portions in the eye of the observer is of a different mixture of wavelengths from the original ray or, by definition, a different color blend. The hue produced by these interfering wavelengths depends on the thickness of the film and the angle at which the ray of light strikes it. If the film is too thick or too thin, interference effects are lost. Moonstone offers a good example of the "schiller," or glow of color produced by interference effects. It contains very thin layers of the mineral albite alternating with very thin layers of the mineral orthoclase. These layers act as films, thus producing the popular bluish and ghostly internal glow by interference when struck by a ray of light. The play of iridescence or tarnish colors on some metals is interference color due to the formation of very thin films of various oxides or sulfides left on the metals by the chemical attack of gases or solutions.

Diffraction: Although it resembles interference color, the peacock play of colors in opal, which is also pseudochromatic, arises from still another process—one which has not been understood until recently, when it became possible to take electron microscope photographs of up to 40,000 magnifications. In these photographs, precious opal is seen to consist of layer upon layer of silica spheres (SiO_2) arranged row upon row in neat, orderly grid patterns with relatively uniform spacings between the spheres. This arrangement acts like the optical-laboratory device called a diffraction grating. This is usually made by scratching a series of fine parallel lines on a glass or metal plate with a diamond point. The lines are spaced as many as 30,000 to an inch. Portions of a light beam directed at such a grating are reflected back from each of the thousands of polished gaps between the scratched lines. Using just one of these tiny "beamlets" as an example, the part that is closest to the edge of a neighboring scratch is bent from its expected path. This bent or diffracted portion is now thrown out of phase with the rest of the beamlet and is in a position to cause interference with its neighboring light waves. The behavior of this single tiny reflection is repeated by all the thousands of others, giving a uniform interference color all across the grating. Precious opal shows its diffraction colors in patches. This is because the grating-like arrangement of silica spheres occurs in irregular patches and the patches are not necessarily oriented in the same direction. The thickness and spacing of

Sapphir.

Rubin.

Spinell.

Smaragd.

Aquamarin.

Alexandrit.

Topas.

Hyacinth.

Chrysolith.

Almandin.

Amethyst.

Demantoid.

43

Below left: Light interference colors are produced by refraction and reflection of light rays by thin films, such as oil, splitting Beam A. Right: To be brilliant, gem must be cut so light entering top will be returned. Opposite: Mid-19th-century microscope, one of earliest for examination of transparent materials under polarized light.

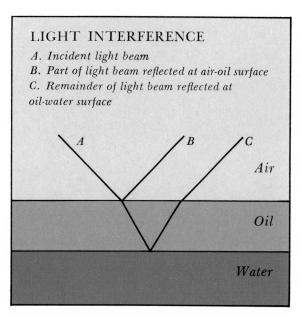

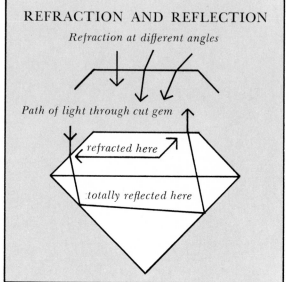

the scratched lines of a diffraction grating have a direct effect on the interference colors produced. The relative positions of the light source, the grating, and the observer also help to determine the colors. So it is with viewing opal. The size and spacing of the silica spheres and the relative positions of the light source, the opal, and the observer make striking differences in the pseudochromatic colors seen.

Refraction: In addition to sifting and sorting and mixing colors or light wavelengths, solid crystal structures can perform other marvelous operations with light. They are even capable of bending it or, perhaps more correctly, changing its direction. All lenses, whether for telescopes or eyeglasses, are designed to take advantage of this fact. By changing the direction of travel of a light image, a lens deceives the eye and brain into giving the impression that the image is coming from a fictitious direction. Light traveling from one sub-

stance to another will be bent varying amounts, depending on the densities of the substances involved. The greater the difference in density the greater the change of direction. Many mineral species can bend or refract light in two directions at the same time. The light beam is actually split into two parts by each part being bent a different amount. Refraction and double refraction are caused by the way a mineral's atoms affect light. Each mineral, then, has its own kind and amount of refraction or double refraction which can be measured and used for identification purposes.

Dispersion: Refraction can also cause interesting color effects. The amount of refraction that takes place depends in good part on what the wavelength of the light is. Blue, for example, is bent more than red. This means that a ray of white light, composed of all colors, by extreme refraction can be separated into its parts and sorted out into a rainbow of colors.

The phenomenon is known as dispersion. Some mineral structures cause greater dispersion than others. The phenomenon is seen almost at its best in diamond which, because of its high dispersive ability, kicks back a dazzling shower of separated color splashes or fire whenever struck by a beam of white light. Rutile and sphene lack most of the other fine gem characteristics of diamond, but they do well in matching its dispersion. Quartz and glass make poor substitutes for diamond because they have so little dispersion. Zircon is a commonly used substitute because it has the fire flashes of high dispersion and hardness and clarity, as well.

Pleochroism: Still another color effect is noticeable among many doubly-refractive gemstones. The beam of light which is split into two beams by double refraction sends its two parts through the structure in different directions. This means that each portion is subjected to different amounts or types of color absorption depending on the direction of passage and vibration. To see this effect, try looking at a light through different directions of the stone. There will be a change in color or in depth of color. The phenomenon is known as dichroism if two color differences can be observed, or by the more general term, pleochroism. It makes a difference, then, which direction the gem cutter follows in preparing cut stones from a rough crystal of tourmaline or ruby or iolite if he wants to take maximum advantage of their pleochroic color differences. Often the color of a tourmaline gem will be entirely too pale if cut lengthwise from a rough crystal, but will be pleasantly colored if cut at right angles to the long direction.

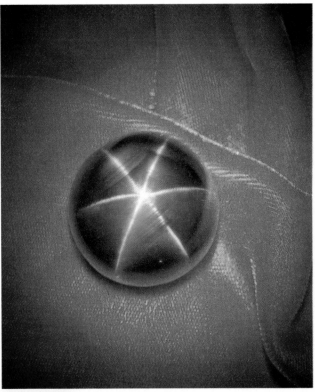

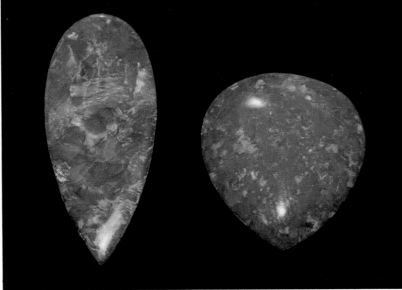

Reactions with Light:
Peristerite from Ontario, Canada
(opposite) is a variety of
plagioclase feldspar
which produces colored schiller
through interference caused
by its thin-plate structure.
Top left: Diamond's brilliant
color flashes are due to
dispersion of white light into
colored parts. Right:
Asterism of star sapphire
—330-carat Star of Asia—is
caused by reflection
from needle-like inclusions.
Left: Color in black opals—44
and 30 carats from Australia—
is diffraction effect.

Transparency and Luster: The two light-reaction phenomena of transparency and luster can almost be thought of as direct opposites. The transparency of a gem is a description of the ease with which light travels through it, while luster is a description of the way in which light is reflected back from it. Transparency is an interesting phenomenon because, except for the evidence of our senses, it seems impossible that anything can pass through a solid substance. And then there is the further problem of explaining how some solids will let the light through and others won't. The fact is that light doesn't go through anything. What happens is that light hits the atoms at the surfaces of these solids and, by its own energy, starts them vibrating sympathetically. These vibrations are passed through the structure from atom to atom. If the atoms are properly aligned, the vibrations will move as in a row of falling dominoes. They are then ejected at the other side in the identical form in which they entered. The light does not trickle slowly through spaces between the atoms as though finding its way through a maze. The rate of travel of the vibrations is close to 670 million miles an hour, which makes its passage seem almost instantaneous.

Luster, being a reflective effect from the surface of the stone, depends on the quality and quantity of the light thrown back. This, in turn, depends on how the stone tends to reflect rather than refract and also on how well the stone can be polished. Both in turn result from the kind of internal structure it has. Most gemstones reflect like ordinary glass and their luster is described as glassy or vitreous. Some, such as zircon and garnet, have a high luster called "adamantine," or diamond-like. A convenient descriptive classification of the kinds of luster might also include resinous (or greasy), pearly, and silky.

Chatoyancy and Asterism: The word "chatoyancy" comes from a contraction of the French words *chat* for cat and *oeil* for eye; it aptly describes this odd reflection phenomenon. The cat's-eye effect is a sharp single band of light running like a brilliant slit across an oval-shaped stone. It looks for all the world like a glowing cat's eye. The light band is multiple reflection from thousands of needle-like inclusions in the gemstone, all running parallel to each other. Interestingly, the thinner and more numerous the inclusions, the sharper and brighter the eye. These needle-like structures may be the mineral species rutile, or may even be extremely thin, empty, capillary tubes. Although the finest cat's-eyes occur in chrysoberyl, they have also been found in some tourmaline, ruby, sapphire, garnet, spinel, and even quartz.

If, by chance, the needle-like inclusions are lined up in two or even three directions related to the crystal structure, the chatoyancy becomes more complex and two or three light bands are reflected. Thus the very popular rubies and sapphires showing asterism, or a star effect, are merely displaying their chatoyancy in three directions with the bands of light intersecting at a single point. Of course, it is necessary to cut such a stone carefully in the proper crystal direction, so that the intersection of the light bands falls at the center of the peak of the rounded gem. Unfortunate-

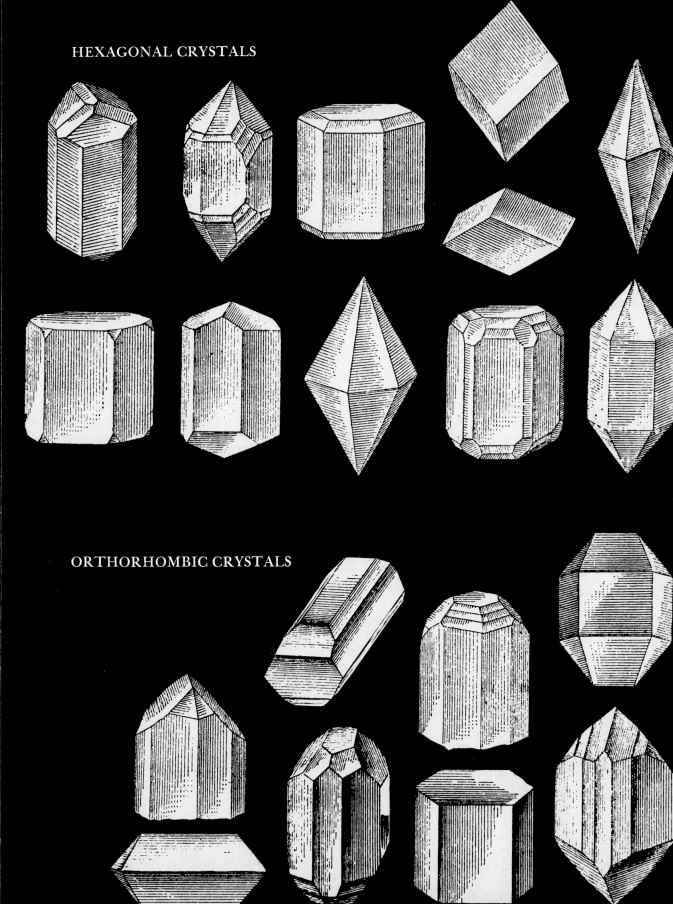

HEXAGONAL CRYSTALS

ORTHORHOMBIC CRYSTALS

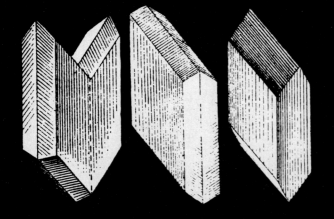

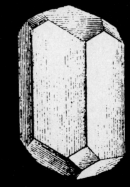

TETRAGONAL CRYSTALS

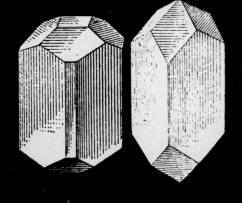

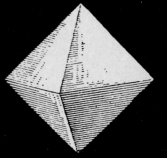

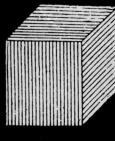

ISOMETRIC CRYSTALS

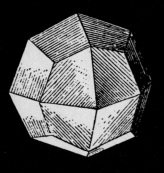

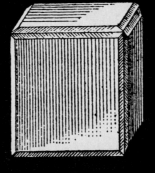

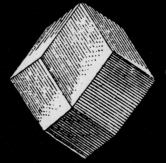

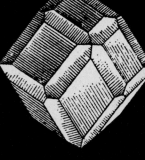

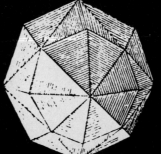

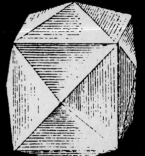

ly, if the inclusions are lacking or more sparse in one crystal direction than in another, a star with weak or missing legs results. All star and cat's-eye stones perform better if viewed under a single point source of light, such as the sun or an incandescent light bulb. Other light sources are likely to be so diffuse as to produce only diffuse reflections.

Luminescence: Most of the light effects discussed so far result from reactions between gemstones and visible light. However, certain gemstones do react to the stimulus of radiation which is beyond the limits of visible light. The most striking effects come from subjecting gemstones to short wavelengths just beyond the violet limits of the light spectrum. This ultraviolet light, or the even shorter X-rays, is absorbed and then given off again as longer and often visible wavelengths, according to a law discovered by Sir George G. Stokes in 1852. Thus, ruby when exposed to an ultraviolet light source will glow like a dull red coal, and some diamonds may assume an eerie, bluish luminescence. The phenomenon is named "fluorescence" after the mineral fluorite in which it was first studied.

Occasionally, the re-emission of this changed shortwave radiation is delayed by the mineral. This phenomenon is known as "phosphorescence." The famous Hope Diamond, when exposed to strong ultraviolet radiation, produces little fluorescent reaction. However, it is startling to see the stone, once the exciting shortwave radiation is removed, glowing with a brilliant scarlet, delayed phosphorescence.

Luminescent effects in gemstones are not particularly important except as identification aids. Unfortunately, a milky blue fluorescence, when it occurs in some diamonds under sunlight or strong incandescent light, can detract considerably from brilliance and value.

Other Physical Characteristics

A gem's effect on light provides many of its attractive features. Other kinds of behavior it may exhibit are just as important and interesting. They, too, are the direct result of the chemical composition and atomic structure of the gemstone species. Among these characteristics are hardness, cleavage, density, and certain electrical properties. For most gemstones, a few of these characteristics—such as hardness, density, and refractive index—are sufficient to make positive identifications.

Hardness: One of the best ways to think of hardness is as the scratch-resisting ability of the gem. Hardness is directly related to the tenacity of atomic attractions. There are great differences in the strengths of the bonds by which different kinds of atoms are held together. Naturally, some combinations will resist being torn apart more than others. The atom bonding in diamond is so very strong that the species is exceptionally hard and cannot be scratched or torn apart by other substances. Softer gems, such as amethyst, are much less strongly bonded and are soft enough so that repeated exposure to scratching forces will leave them badly marked. Since the early 1800's a rough but convenient scale for measuring hardness, originated by the German mineralogist Friedrich Mohs, has been in general use. The scale is based on ten relatively common minerals ranked from 1 to 10 in the order of increasing hardness: 1) talc, 2) gypsum, 3) calcite, 4) fluorite, 5) apatite, 6) feldspar, 7) quartz, 8) topaz, 9) corundum, 10) diamond. The degree to which hardness increases between the numbers is not at all uniform. There is a greater difference between the hardness of corundum and diamond—9 and 10 in the scale—than between numbers 1 and 9. Almost all important gemstones have a hardness above 6 in this scale. Anything less than 6 is not durable enough to resist the scratching and chipping of general use. For practical purposes it is useful to know that window glass is usually slightly softer than 6, a good steel knife is 6 to 6½, and a hard file is close to 7. One of the odd variations in the hardness of some gemstones occurs when there is an appreciable difference in the strengths of the atomic bonds in different directions through the structure. Kyanite, for example, varies from 5 to 7 in hardness, depending on the direction in which an attempt is made to scratch it.

Cleavage: Gemstones do break. Some come apart with relative ease, while others require the shock of an energetic blow or the unusual stresses that result from sudden and extreme cooling or heating. The ease with which they break up depends upon the strengths of the bonds between atoms holding them together. If the breakage is irregular, leaving uneven or jagged surfaces at the break, it is called "fracture." At times, with different minerals, the breakage will follow definite directions. The crystal structure splits where the bonding is weakest between certain planes of atoms in the internal arrangement. This causes the broken surfaces to be flat faces parallel to planes of atoms which are parallel in turn to possible crystal faces. Topaz, for example, always breaks in a manner showing flat surfaces or cleavage planes that develop in one same direction through the structure. Other minerals may exhibit cleavage in two, three, or even four different directions through the crystal structure. Feldspar cleaves in two directions which are at right angles to each other. Spodumene also has a two-directional cleavage, but the two are not quite at right angles. Since the cleavages of different kinds of gems vary considerably, they can be used to help identify the stones.

Weight and Specific Gravity: Contrary to popular understanding, the fact that a gem weighs, say, ten carats has little to do with its size, except when compared with a gem of its own species and of similar cut. The carat is an expression of weight, not size. One carat is equal to one-fifth of a gram; in more familiar units, there are about 140 carats in an ounce. This standard unit of measurement of gemstone weights was not legalized in the United States until 1913. Before 1900 the several weight units used throughout the world differed somewhat, but the carat in one form or another had been used since ancient times. Speculation relates the weight of the carat to the weight of a seed of the tropical carob tree. These seeds are unusually constant in weight and are just slightly lighter than our legal metric carat. The ancient Greek weight—the ceratium—and the Roman siliqua are both about the same as the carob seed. Most likely, then, the name and the weight did originate with the carob tree.

Even knowing that the carat is an expression of weight only, gem buyers are often surprised to see that a one-carat sapphire is considerably smaller than a one-carat diamond. Sapphire is denser than diamond and a smaller stone can weigh the same number of carats as a larger diamond. Since size and weight are both very important measures for gemstones, the unit of measurement that combines both—"density"—is fundamental. It is complicated to think of a gem's density, because size and weight must be considered simultaneously. We are aware of a difference in weight when we compare iron and wood. Yet it would not always be correct to say that iron weighs more than wood because a large piece of wood can weigh more than a small piece of iron. Only by comparing equal volumes of these materials can the extent of the weight difference be made clear and unmistakable. Making innumerable comparisons of the weights and volumes of different solids with each other seems inconvenient. To avoid this, it is customary to compare all of them with some selected substance, readily available, with a known density. Water is the standard most often used. Ruby, for example, proves to be four times heavier than an equal volume of water, so its specific gravity—the term of comparison with water—is 4. Diamond weighs 3½ times as much as an equal volume of water and thus has a specific gravity of 3½.

Electrical Properties: Gems may have a number of other interesting characteristics unrelated to their appearance or durability. Among these, electrical behavior is sometimes remarkable. Museum curators and jewelers have long known that their tourmaline specimens and gems will accumulate thick coats of fine dust in a short period of time even when they are displayed in tightly sealed showcases. As the case lights are turned on and off each day the tourmaline is alternately heated and cooled. When heated, tourmaline develops a substantial electric charge which quickly attracts tiny dust particles in the air. Before long, a gem will be coated. This characteristic is known as pyroelectricity—electricity produced by heating.

Diamond, topaz, tourmaline, and amber, when polished briskly with a cloth, will even develop enough of an electric charge to attract and hold small bits of paper.

An electrical effect discovered by Pierre and Jacques Curie in France in 1880 is perhaps even more remarkable. These two physicists, studying the ability of crystals to conduct electricity, found that when certain crystals were squeezed they developed a measurable electrical charge. The phenomenon was later called "piezoelectricity," based on a Greek word *piezin,* meaning "to press." As early as 1881 another Frenchman, G. Lippman, predicted that a reverse phenomenon would take place. He suggested that an electric charge placed on any piezoelectric crystal would cause it to change shape. This was successfully tested by the Curie brothers. Today this important characteristic is applied, using quartz and other piezoelectric substances, to control the frequencies of all radio broadcasting and other electronic devices. The alternating current in these devices and the piezoelectric crystal plates inserted in them must operate in unison. Thus,

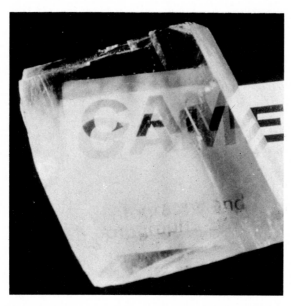

the fixed dimensions and structure of the plate keep the current alternating only at the rate at which the crystal plate can change its shape. Such a device is known as a frequency control oscillator. It is responsible for the fact that every time you tune in your favorite radio or television station it is at the same number on the dial, just where it was last time.

Finding Gem Characteristics

The gemologist often is handicapped in trying to determine gem characteristics. If the stone is mounted in some kind of jewelry setting, manipulation for tests is difficult. Owners frequently object to removing stones from their mountings since damage is possible. Even if the stone is not mounted, it is not practical to risk ruining the cut or polish by scratching, chipping, or removing a piece for testing. Every bit of damage reduces the value of the gem. Testing on cut stones is usually

limited to nondestructive manipulation. Unfortunately, some of the best gem study techniques involve destruction of the sample. Destructive tests are carried out, then, on uncut gemstones. Among such tests are chemical analysis, X-ray structure determination, and Mohs' tests for hardness. Nondestructive tests include determination of refractive index, specific gravity, pleochroism, spectral pattern, and examination for any foreign inclusions in the stone.

Hardness: As already mentioned, hardness testing is destructive or damaging to the stone. It is determined by actually trying to scratch the stone with some of the minerals in Mohs' scale of hardness. Bits of the harder minerals in the scale—from 5 (apatite) to 10 (diamond)—can be obtained already mounted in small metal rods for convenient manipulation. Carefully, an attempt is made to scratch the gem with one of these hardness pencils, perhaps #7. The scratching is done under magnification and along the edge of the gem where it will not mar any of the facets. Only a tiny, almost invisible scratch is necessary. If #7 will not produce a scratch, #8 is tried. Should #8 produce a scratch then it is obvious that the gem's hardness lies between #7 and #8. Estimates can be made about whether it is 7¼, 7½, or 7¾, depending on how easily #8 made the scratch.

Refractive Index: It has already been explained how minerals refract or bend a beam of light. As the light hits the flat surface at an angle, it bends upon entering the gem. If the light beam's direction is slowly changed so that it comes to the surface of the gem at an increasingly lower angle, eventually a point is reached

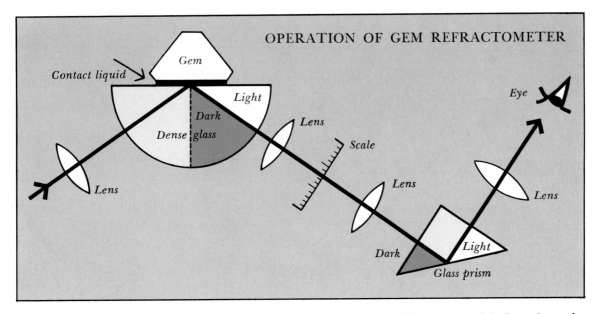

OPERATION OF GEM REFRACTOMETER

Contact liquid

Gem

Light

Dark

Dense glass

Lens

Lens

Scale

Lens

Dark

Light

Glass prism

Lens

Eye

where it ceases to bend sufficiently to enter the gem. It just grazes the surface. Any further lowering of the beam causes it to be totally reflected away from the gem. The grazing angle is called the critical angle, and it will differ with each gem substance according to its refracting ability. An instrument, the gem refractometer, has been devised to measure this critical angle quickly and easily. The instrument usually contains a built-in scale from which the refracting ability, or "refractive index," of the gem can be read directly. To read the index on a typical gem refractometer, of which there are several models on the market, one of the polished faces of the gem is placed against a polished piece of very highly refracting glass mounted in the instrument. Good contact is assured by placing a drop of highly refracting liquid between them. A light beam is brought through the glass to the gem. Any of the light coming to the gem from an angle

at which it will be refracted is bent into the gem, away from the instrument, and is lost. Light coming in at an angle beyond the critical is reflected back into the instrument, hits the viewing eyepiece, and its trace shows as a bright section on the scale. The numerical marking on the scale dividing the light portion—representing reflected light—and the dark portion—representing the lost light refracted into the gem—is the critical angle. For convenience, this is numbered on the scale as the refractive index. The measurement is sufficiently precise so that, by consulting a table listing the refractive indices of gemstones, one can usually make a quick identification of the gem in question.

There are other methods for determining refractive index that are somewhat less convenient. One of the best of these uses a series of fluids of known refractive indices. When a gem is placed in a liquid having the same

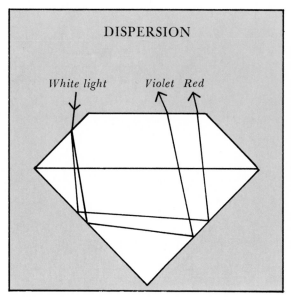

DISPERSION

White light   Violet Red

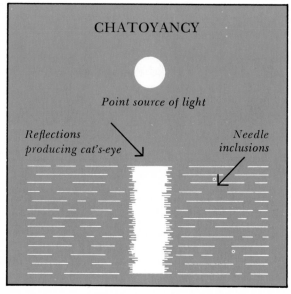

CHATOYANCY

Point source of light

Reflections producing cat's-eye   Needle inclusions

refractive index it effectively disappears. This happens because the gem does not cause any further bending of light than that already done by the liquid. We have no visual evidence of the gem's presence because it behaves toward light exactly as the liquid does. An appropriate series of test liquids such as clove oil (index 1.54) to test for quartz (index 1.54), cassia oil (index 1.60) to test for topaz (index 1.61) and methylene iodide (index 1.74) to test for spinel (index 1.72-1.73) is easily assembled. Although some of the liquids are expensive and difficult to obtain and may be unstable enough to need frequent replacement, the method has some benefits. It can be used with very small gemstones fragments, and the flat, polished surfaces required for the refractometer are not necessary. Also, even when the gem material disappears in a liquid of appropriate index, the cracks, flaws, and foreign inclusions in it do not vanish. This affords an excellent method of

checking a gemstone internally to see how it can best be cut to make the most perfect gem. It also may make it easier to see and study the nature of the inclusions.

Specific Gravity: Most gems and gemstone samples are large enough to be weighed accurately, so that one of the weight-measurement methods is normally used to find the specific gravity. You will remember that specific gravity is the weight of the gem compared with the weight of an equal volume of water. The Greek mathematician Archimedes is credited with having discovered the method of determining specific gravity in the third century B.C. He realized that an object would weigh less when submerged in water and the weight loss would be equal to the weight of the water whose place was taken up by the object. In other words, the weight loss represented the weight of a volume of water equal to the volume of the object. All that is necessary to determine specific gravity

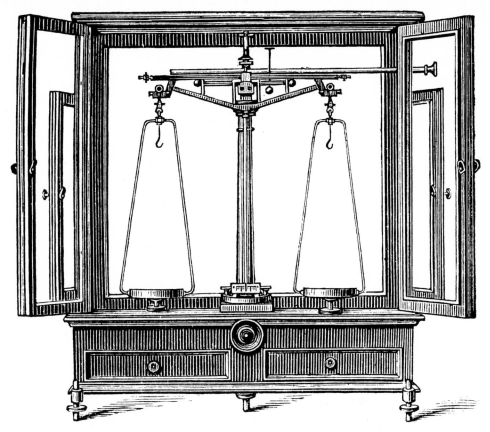

is to weigh the gem accurately in air and weigh it again while it is immersed in water. A simple calculation gives the answer. The gem's weight in water is subtracted from its weight in air. This gives the weight of a volume of water equal to the volume of the gem. This weight is divided into the weight of the gem in air to find how many times it exceeds the weight of the water—thus its specific gravity.

Of course, the more accurate the scale for weighing the more precise the determination of specific gravity. For larger gems and gemstone fragments, cruder and less sensitive weighing devices—even homemade—are accurate enough.

As with refractive index, it is also possible to find the specific gravity of a gem or fragments of a gemstone by using a series of liquids. It is appropriately called the "sink-float" method. In this procedure the liquids have a known specific gravity. Very simply, if a stone is denser than a liquid it will sink, if less dense it will float. A more precise measurement can be obtained by first floating the gem on a denser liquid. Slowly, drop by drop, with stirring, a less dense liquid is added. Eventually, the gem will start to sink as the mixture reaches a density just slightly less than its own. Quickly, before evaporation can cause changes, the density of the liquid mix is determined; it will be equal to the density of the gem. Of course, this requires the availability of a floating instrument called a hydrometer. This floats in the liquid and the density can be taken from a number scale which is read at the mark where it settles at the surface. There are also other, more accurate and more expensive kinds of weighing devices for finding the density of such mixed liquids.

Pleochroism: As explained before, different colors can often be observed in double refracting gems by looking at them from differ-

57

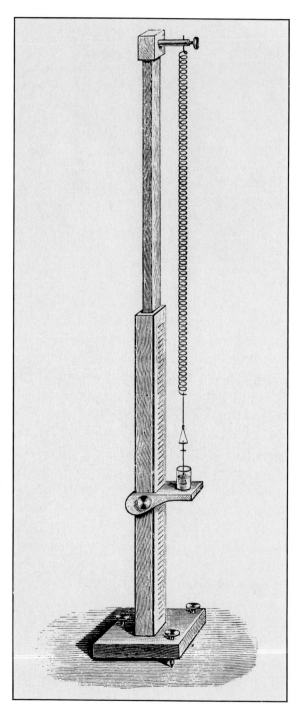

ent directions. Any color differences seen must be remembered. This dichroism or pleochroism can be seen, and the colors compared directly, without the necessity of relying on memory, by using a dichroscope. In its simplest form, this instrument uses two small squares of Polaroid sheet which are fastened with their edges touching but with their polarizing directions set at right angles to each other. The gem is viewed through the Polaroid against a strong light. If dichroism is present, the color of the gem portion seen through one section of the Polaroid will differ from that portion seen through the other. By turning the gem in several directions, a third color may possibly appear. If there are observable color differences, the gemstone is doubly refracting and cannot be an isometric mineral or glass or plastic. If only two colors are visible, very likely the mineral is tetragonal or hexagonal. Three colors will almost guarantee that the mineral is orthorhombic, monoclinic, or triclinic.

Sometimes dichroism and pleochroism are a little difficult to see because the color differences may be subtle enough to escape visual detection. By itself, the use of the dichroscope is not a positive means of identifying gemstones, but it does add helpful information to that obtained by other tests.

Spectrum Analysis: For colored gemstones it is often possible to obtain very useful information for identification by use of the gem spectroscope. The instrument's operation is based on the separation of white light into its complete rainbow, or spectrum, of colors. This is done by a built-in prism which receives the

58

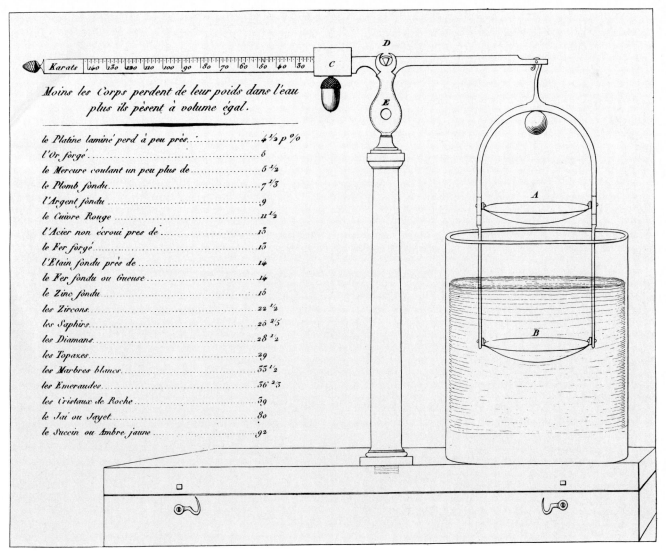

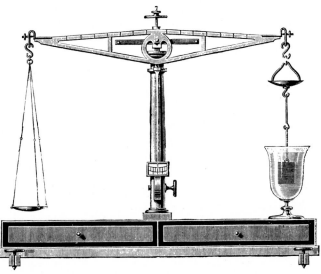

Three Designs for Specific Gravity Balances: Jolly type (opposite) shows weight changes on vertical scale as spring varies. Above: single-pan, beam-type balance uses counterweight and horizontal scale. Left: Double-pan beam type achieves balance by adjusting counterweights in left pan.

light through a narrow slit. The prism sorts out the various wavelengths by its strong dispersion. Often a diffraction grating is used instead of a prism to diffract or separate the colors. Looking through the opposite end of the instrument one can see the continuous rainbow as a band of touching parallel bars or lines of different colors. They range from violet and indigo at one end, through blue, green, yellow, orange, to red at the other end. Now the colored gem is placed between the light source and the spectroscope slit. As expected, certain specific colors from the white light are absorbed by the gem and do not enter the spectroscope. Their absence causes black bars—the absence of specific colors—to appear in the continuous spectrum of color bars. For some colored gems these patterns of black bars are very distinctive and are good identification features.

Gemstone Inclusions: The student gemologist, when he first begins to look inside gemstones under magnification is often amazed at the myriad of included objects he sees there. Tadpole, comma, round, or elliptical bubbles abound as well as fibrous horsetails, cracks, beautiful tiny crystals, blades, feathers and mossy traces of matter. At magnifications of from 10 to 40 times under the microscope it is often possible to recognize inclusions that tell not only what the gem is but where in the world it came from. Tiny actinolite blades in an emerald signal immediately that it was mined in Russia and not at the famous mines of Colombia.

These inclusions may have arrived in the gemstone at different times. Some existed before the gemstone was formed and they were swept up and trapped in the developing solid. Even tiny droplets of the liquid from which the crystal formed are sometimes trapped. Some liquid-filled inclusion cavities have tiny gas bubbles that move back and forth in their small prisons as the stone is tilted. Now and then, a mineral species developing from the same liquid as the gemstone leaves its trace as a scatter of bright, little but well-formed crystals peppered through the stone. Often, too, after the gemstone has formed it develops a series of tiny cracks and fissures. These may later be filled by the infiltration of liquids which form new crystalline material to "heal" the breakage. This accounts for the typical "healed" feathers seen in Ceylon sapphires. There are certain significant internal features caused by accidents during growth. Color zoning, sometimes not too obvious without magnification, will appear as definite bands of differing color intensity due to interruptions during growth or slight changes in the content of the supply of material brought to the forming crystal. Prominent hexagonal color zoning is typical of Burmese sapphire.

Much work has been done toward identifying and classifying the various kinds of inclusions to assist in the identification of gemstones. It has been found that, in general, inclusions fall into four major groups of characteristics: (1) zoning and spotty distribution of color, as well as other markings caused by accidents of growth; (2) solid particles of a different mineral species, or even separate crystal pieces of the same species; (3) cracks and other openings either "healed" by the addition of new substances or filled with gas or liquid; (4) cavities holding gases, liquids, or small solid crystals and sometimes all three.

X-Ray Examination: Although the kind of X-ray examination of most value to gemology requires the destruction of a small amount of the stone it remains the most important single key to identification. When a beam of X-rays is directed at a solid, much of it passes through the solid with no alteration, some of it is scattered, and some is converted to heat or other kinds of energy. It is the scattered X-rays that are of interest because they are the ones that have hit atoms on the way through. A picture of these scattered X-rays is taken by proper placement of a photographic plate. A large number of atoms, all uniformly spaced and placed in the structure, will scatter the X-rays in the same direction and reinforce their image on the film. The others are scattered in various directions and produce no combined mark on the film. This means that every different plane of atoms in the structure leaves its print on the film. By careful measurement of the markings on the film and suitable mathematical treatment of the measurements, the entire internal atomic structure is revealed.

The X-ray method most often used is the powder diffraction procedure. The camera is a flat hollow metal cylinder, one end of which is a removable lid. It is very carefully machined to exact dimensions, since its uniformity and the size of its diameter are crucial in the final film measurement. There is a hole on one curved side for the entrance of X-rays, and a hole opposite for the exit of most of them. Tapered metal tubes fit in these entrance and exit ports to guide the X-rays to and from the sample. The sample is mounted on a rotating spindle at the exact center of the camera. The film is a long, narrow strip that fits flat against the inside wall of the cylinder. It has two holes that fit over the entrance and exit port tubes.

In operation, the sample—a tiny bit of the mineral which has been powdered and then held together by various adhesives—is mounted, the film is loaded in the darkroom, and the lid is replaced. As the sample spindle is turned by a motor-driven belt, the entrance port of the camera is placed against the X-ray source and the exposure proceeds, taking several hours. When the film is developed, it has a series of matched curved lines running across it. These lines represent atomic planes; for each species their pattern is characteristic and will be different from that of any other species. The films can be indexed and filed and used much the same way as fingerprints.

All told, then, in their investigation and study of gemstone species and gems through the years, mineralogists and gemologists have assembled a rather impressive arsenal of instruments and techniques. The accumulation of facts has also proceeded steadily until we have reached a point where the many problems of gemstone identification and gem preparation are matched by sufficient knowledge to solve them. It is a very rare or unusual natural gem material that worries a qualified and competent gemologist. However, man is also perfecting his ability to manufacture gemstones. It is obvious then that the science of natural gemstones is essential if the distinctions between natural and man-made gems are not to be obscured.

Treated Gems & Substitutes|3

Without question, there are far more substitute and make-believe gemstones in circulation throughout the world than there are natural, properly identified stones. This is an ancient state of affairs persisting since the beginning of man's experience with gems. Pliny's *Natural History* summed up the situation beautifully for all time. "Truthfully," he wrote, "there is no fraud or deceit in the world which produces larger gain and profit than that of counterfeiting gems." Very likely there are more profitable endeavors and very likely the statement is a little strong and uncharitable toward those engaged in this trade. Most frequently gem counterfeiting is not intended as fraud. When fraud occurs it is not the fault of the materials or their producers necessarily, but rather the motives of the buyer or seller. Gem counterfeiting caters to a very large market composed of those thousands of people who feel they cannot afford the price of an attractive natural gem. Also, there are many who appreciate the decorative value of gems, but prefer to make their investments elsewhere.

Unfortunately, the existence of a large, readily available supply of counterfeits does cause some confusion in the gem markets. It is undoubtedly responsible for some of the strong resistance to buying the more expensive, non-counterfeit gems. The resistance stems from a feeling among potential buyers that they are at the mercy of the jeweler. It is true that with little knowledge of the subject and without the ability to distinguish between the real and the unreal gem, most buyers must rely on the integrity and knowledge of the jeweler.

The gem-buying public does feel that there are detectable differences between the real and the counterfeit; it feels that experts can detect these differences easily. As long as these beliefs and feelings are substantially supported by the facts, there will be a strong market for the real, natural stones. At the moment, most problems in detecting counterfeits are rather easily solved. However, as new techniques for manufacturing or tampering with gemstones develop, identifications become more complex and differences more difficult to detect. It doesn't make a gemologist feel too comfortable to admit that his first suspicions about a questionable ruby were aroused because the gem was entirely too beautiful and perfect to be natural.

The controversy over gem counterfeiting is not a simple one of real gems versus counterfeits. There are all sorts of counterfeits, shading from those almost completely identical to their natural counterparts across the scale to cheap glass and plastic imitations that need no expertise to be detected. The simplest gem counterfeiting is done by substituting a less expensive gem for one of higher value it closely resembles. In addition, there are imitation gems, assembled gems, reconstructed gems, manufactured gems, and even treated and altered natural stones.

Look-alike Gems

Superficially, certain gems resemble others—at least in color. This suggests that a lesser gem may possibly be used to play the part of a more expensive cousin. Usually, such substitutes are satisfactory at a distance but give themselves away under close scrutiny. A few are deceptive enough to need expert measurement before the masquerade can be detected. Diamond is a common target for mimicry because it is expensive and so much admired. The most successful gem substitutes for it are colorless zircon and colorless sapphire. Colorless quartz has also been used but with less success. Zircon is best because it has higher dispersive ability than the others and is able to produce a respectable amount of fire. Sapphire, quartz, and glass have low dispersion with resulting low fire. In hardness, zircon barely holds its own at 7 and it also chips easily on the edges. Colorless sapphire has a hardness advantage at 9, but without fire it is easily detected. Zircon, sapphire, and quartz are all doubly refracting stones and diamond is singly refracting. Looking through one of these substitutes at the edges of the back facets shows, under magnification, an apparent doubling of these edges due to the doubly refracted image. In contrast, with diamond each edge appears as a single line.

Red garnet and red spinel make good substitutes for ruby. Detection in this case is often difficult except by close measurement and observation. Undoubtedly, "rubies" now gracing all sorts of jewelry of any vintage have a good chance of not being rubies at all. Good red rubies are very rare. The Black Prince's Ruby and the Timur Ruby in the British crown jewels have already been mentioned as prime examples of red spinel masquerading successfully as true ruby. Some combination of measurements of specific gravity, hardness, and re-

fractive index can establish the truth. Unfortunately, such measurements can't be made easily on gems mounted in jewelry. Microscopic examination of tell-tale inclusions may be enough to substitute for other tests. The "silk" often found in Burmese rubies is a good example. "Silk" is the gemologists' name for the shimmering, whitish sheen caused by light reflection from a series of many short, tiny needles of rutile arranged at angles of 60 and 120 degrees to each other following the hexagonal structure of the ruby crystal. Ceylon rubies also may show rutile inclusions, but they will be longer needles more widely spaced. They are also liable to contain tiny brown zircon crystals. The extensive study and photography of gem inclusions done by E. Gubelin in Switzerland, and others, has supplied valuable information for the detection of substitutes and misidentified gems.

The list of gem-for-gem substitutes is much longer than these few examples. It should include certain colors of green tourmaline substituting for emerald, or even yellowish-brown citrine substituting for topaz. This replacement is further complicated because citrine—a yellow-brown variety of quartz—often goes under the name "Brazilian topaz" in the gem trade. The worst offenders of this sort are sapphires of various colors. They have been marketed as "oriental topaz," "oriental emerald," "oriental aquamarine," and "oriental amethyst," depending on the gem they resemble and might possibly replace. As numerous as these substitutions may be, they can be correctly identified one way or another, and usually with ease.

Imitation Gems

The Egyptians are given credit for the first successful attempt to manufacture imitation gem materials. Recovery of objects dating as far back as 4700 B.C. gives solid evidence of their gem-making advances. Beads and pendants were made at that time of a substance called faience—an inner core of powdered quartz grit covered with a layer of colored, glass-like glaze. Shortly after 1600 B.C., true glass was used in Egypt for the same purposes as faience. Now, some thirty-six centuries later, glass in its various forms is still the most commonly used material for gem imitation. There are as many formulas, colors, and general characteristics for glass available to the present-day gem maker as he could possibly want. All glasses, however, lack one crucial characteristic—hardness. Continued use will quickly mar, scratch, and dull imitation gems of glass. Even

so, many things can be done to glass gems or "paste" to make them attractive, and when they deteriorate the cost of replacing them is slight.

Basically, glass is formed by melting together quartz sand, soda (sodium carbonate) or potash (potassium carbonate), lime (calcium oxide), and cullet, or waste glass, to help the melting process. Other substances can be added, such as lead oxide and barium oxide to increase brilliance and weight, metallic oxides of various kinds to give color, and manganese oxide to take out any color. Typical coloring oxides are cobalt oxide for blue and chromium oxide for green.

The batch is heated in a crucible and coloring oxides added. When the constituents are melted together and stirred into a homogeneous mass of liquid, the batch is cooled. Since cooling is too rapid for crystallization to occur the mass only seems to become solid. Ac-

tually, it is a supercooled liquid with its atoms disorganized. Glass, if it sits long enough, will eventually organize its atoms and crystallize, thereby losing its glassy characteristics. Since glass is of variable composition, has no crystal structure, and can be any color, it will not have stable characteristics by which identification can be made.

The best quality glass gems are made of "strass," which varies a bit in composition but is basically a melt of quartz, red lead oxide, and potassium carbonate. It is a brilliant, soft glass with a relatively high index of refraction. This glass was named after the Austrian, Josef Strass, who is acknowledged as the first to use it. Such glasses, containing lead oxide, are classed as "flint" glasses. Because of their higher refractive indices they are very good for optical purposes. The cheaper "crown" or bottle glasses are used for making gems for costume jewelry. Whatever the glass type, there are several ways

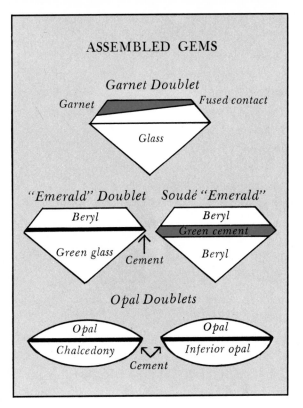

ASSEMBLED GEMS

Garnet Doublet

Garnet   Fused contact

Glass

"Emerald" Doublet

Beryl

Green glass   Cement

Soudé "Emerald"

Beryl
Green cement
Beryl

Opal Doublets

Opal
Chalcedony

Opal
Inferior opal

Cement

67

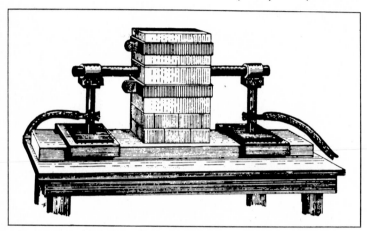

of finishing off the gems. The glass, still molten, may be placed in a mold. When cool, the gems are broken out and used directly, or they may be polished on the top face. Better gems are polished on all faces. In some cases, the gems are actually cut to order from a glass rod or sheet.

For various reasons—usually to improve the product or to cut costs—the glass imitations may be treated. For example, to avoid stocking many colors, producers keep a quantity of colorless paste gems and when orders for particular colors come in, they spray the backs of the colorless gems with a pigment and cover this with a mirror surface. Sometimes just the mirror back is applied to increase brilliance. An older but similar practice was to put a piece of foil behind the paste gem in a sealed jewelry mounting.

Special opaque or semiopaque glasses have also been devised for gem substitutes. Lapis lazuli, jade, and turquoise have all been successfully imitated in glass. Aventurine glass

or goldstone resembles no natural gemstone but has become popular in its own right for costume jewelry. It is a crown glass to which copper oxide has been added. As the glass forms, tiny hexagonal plates of copper metal form as a suspension in the mass. They give it a golden sparkly appearance as light flashes reflect from a multitude of platelets.

In recent years an enormous array of plastics has become available for gem imitation. Because plastics are so soft and lack brilliance they are usually less satisfactory than glass. However, for decorating fabrics they are more practical, as they reduce weight significantly. Plastics will also reproduce faithfully the outlines of a mold and, unlike glass, will form gems with sharper edges, more like hand-cut stones. The most confusing plastic imitations are of Bakelite and other resins which simulate amber. They can be distinguished from true amber but often only by sensitive chemical testing. Celluloid was the first plastic to make its way into the market for gemstone imita-

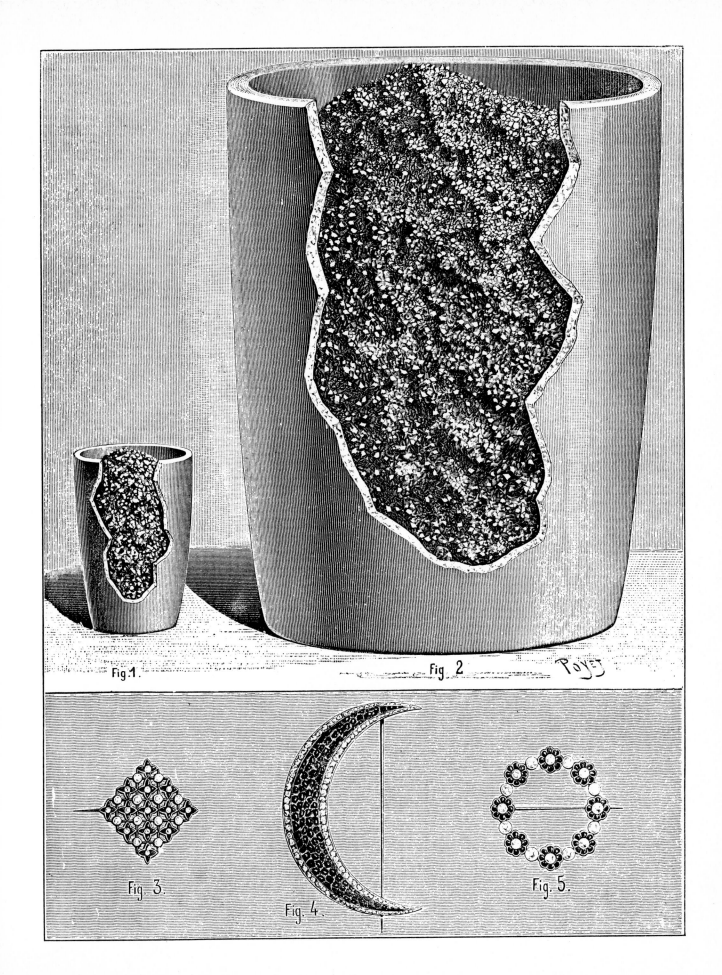

Fig. 1.

Fig. 2.

Poyet

Fig. 3.

Fig. 4.

Fig. 5.

French chemist Auguste
Victor Louis Verneuil and his
diagram of first flame-fusion furnace
(published in 1904) with which
he made synthetic ruby and
sapphire. Below is simplified
diagram of basic parts.

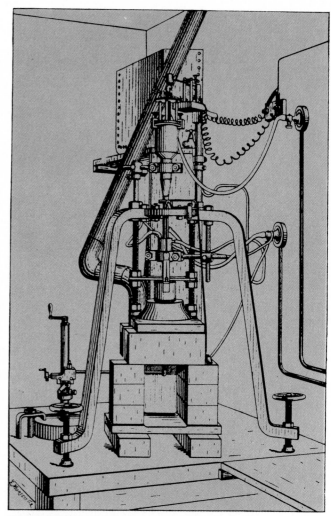

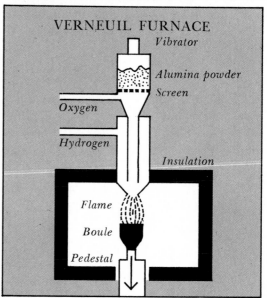

VERNEUIL FURNACE

Vibrator

Alumina powder

Screen

Oxygen

Hydrogen

Insulation

Flame

Boule

Pedestal

tions. Quantities of carved "ivory" were produced from it. Many of these carvings were only molded but some were actually carved. Unfortunately, the substance is almost explosively inflammable and it has been replaced by cellulose acetate or, more recently, by other plastics containing white filler and pigments.

Both glass and plastics are materials so far removed from the traditional gem requirements of durability and rarity that there is little danger they will upset the demand for true gems. The gem market is, after all, a market in luxury goods, so that appearance, economy, and availability have not and will not erode the desire for something durable and at the same time rare.

Assembled Gems

In an attempt to improve the quality of gem imitations, many ingenious kinds of composite gems have been devised. By cementing or fusing together pieces of imitation or natural gem material, a more acceptable—and in some cases more deceptive—gem may be produced. The simplest of these composite gems is called a "doublet," in reference to the two-piece assembly. Since a major problem with imitation gems is lack of hardness, a cap of some harder material helps. A typical doublet is made by fusing a thin plate of garnet to a colored glass blob. The assemblage is then cut and polished so that the garnet portion comes at the table, or top, of the gem. With such a thin plate the red garnet color doesn't show, but the entire gem assumes the color of the glass bead. This kind of doublet is easily detected, but others may be more difficult and thereby lend themselves to

fraud. For example, a colored bottom of green glass may be cemented to a top of colorless or weakly colored beryl. If hardness and refractive index are checked, as is usual, the top of the gem shows the correct measurements for beryl, while the green glass color dominating the stone makes it look like bright green beryl —or emerald. A diamond top on a colorless sapphire or colorless spinel base may be mistaken for a complete diamond. Any of these doublets is far from obvious if the joint between top and bottom is carefully concealed in a jewelry mounting. A most successful counterfeit is the *soudé* emerald, made top and bottom of pale beryl, quartz, colorless sapphire or colorless spinel, with its parts joined by a thin layer of green cement which gives color to the entire stone.

Opal lends itself very handily to the manufacture of doublets. Most opal is opaque, so that it is next to impossible to see the cementing plane through the top of a doublet. As long as the joint around the edges is covered by a mounting even the bottom may be safely exposed. If the bottom piece is made of very poor quality opal this is acceptable because, even with the best of opals, the back is never expected to equal the front in quality. Other kinds of opal doublets have been made. Sometimes a backing of black glass or other black stone may improve the color display of a thin top of precious opal. A doublet of sorts—imitating opal with limited success—has been made from a top of glass or colorless quartz cemented to a plate cut from iridescent abalone shell. Ingenuity has even led to the production of an opal triplet. A typical opal doublet is capped

71

*Diagrams of apparatus for flux
(left) and Czochralski methods of growing
synthetic crystals. Opposite: Gem crystal
of YAG (yttrium aluminum garnet),
a light-colored, fairly hard (8), recent
synthetic with high refractive index.*

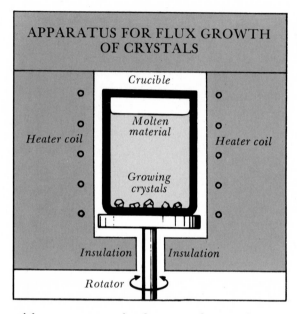

APPARATUS FOR FLUX GROWTH
OF CRYSTALS

Crucible

Molten
material

Heater coil   Heater coil

Growing
crystals

Insulation   Insulation

Rotator

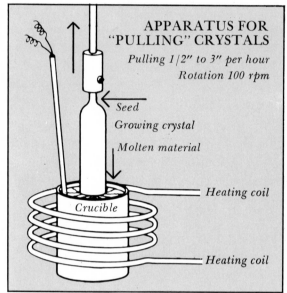

APPARATUS FOR
"PULLING" CRYSTALS

*Pulling 1/2" to 3" per hour
Rotation 100 rpm*

Seed
Growing crystal
Molten material

Heating coil

Crucible

Heating coil

with an appropriately curved covering of quartz which improves the brilliance of the color display while protecting the softer opal from scratching, wear, and shock.

Triplets are not limited to opals. Occasionally the soudé emeralds already mentioned will have a cemented center layer of green glass sandwiched between the top and bottom pieces, making them technically triplets. Poorly colored jade cabochon doublets have been made with hollow centers into which small pieces of jade are cemented with bright green cement. These too are triplets with the bright green jade color coming from the cementing material.

Deceptive as they are under cursory examination, assembled stones are quite easy to detect under low magnification. Usually the joints between pieces are obvious unless artfully concealed. Even then, by looking into the

stone from the top a single layer of air bubbles can be seen at the cemented or fused joint between the parts. Often, too, the inclusions in the top and bottom pieces are markedly different and never cross over from top to bottom.

Reconstructed Gems

The only successful attempts at reconstituting small, valueless gem fragments into marketable gems have been made with ruby, beryl, and amber. In the early 1880's in Switzerland an unknown counterfeiter produced some rather respectable rubies in his laboratory. When these were sold through the gem trade, trouble quickly developed. It was discovered that they had been prepared by melting together small fragments of natural ruby. Experimentation soon showed that ruby grains, heated to about 1800 degrees centigrade would soften. At this stage the grains welded together and could be

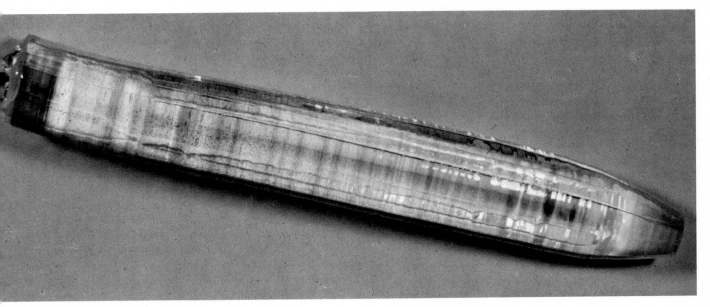

cooled to a single solid mass. Unfortunately, at 1800 degrees centigrade the chromium coloring agent is driven off, leaving a pale to colorless mass. Previously, in 1837, a Frenchman, Marc A. A. Gaudin, had reported that green chromic oxide added to aluminum oxide and heated would produce the ruby color. All it took to restore the color was the addition of chromic oxide to the reconstituted ruby before cooling. Once the true nature of the reconstructed "Geneva rubies" was discovered the bottom dropped out of the market and they have not been made commercially since the early 1900's.

Reconstructed amber has been more persistently successful. In the recovery and mining of amber, quantities of gem-quality pieces are obtained that are too small for cutting or carving. In a hydraulic press at high pressure and at about 180 degrees centigrade—in the absence of air—these amber fragments are pressed and flow together to form ambroid. It has both the appearance and characteristics of untreated amber.

Attempts to melt or fuse together small fragments of emerald, or other colors of beryl, have produced unexpected results. The beryl after heating and cooling results in a true glass being formed with a beryl composition. This glass has a lower density, lesser hardness, and lower refractive index than true crystalline beryl. Therefore, even though it can be appropriately colored, it does not have the desired beryl characteristics and makes only a fair substitute.

Manufactured Gems

The subject of man-made gems is extremely important in any discussion of gems. By far the best substitutes are the manufactured or syn-

73

74

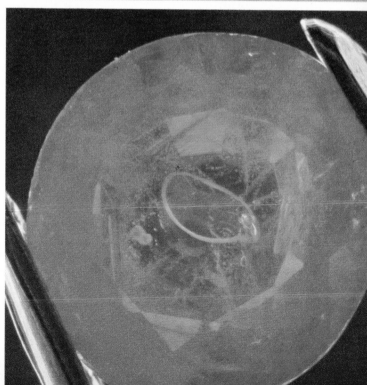

thetic gems. It is sometimes difficult even to think of them as substitutes because they are usually identical in chemical composition and physical characteristics with their natural counterparts. In fact, they frequently surpass natural forms in size and perfection. A ruby of ten carats with a minimum of inclusions or flaws is a rare thing in nature. Flawless, fine colored rubies of almost any desired size are synthesized in industry in enormous quantities. Price differences between natural and synthetic rubies of near equal perfection are ridiculously great, the natural ruby costing perhaps five hundred times as much.

The great breakthrough, when man found a way to make fine gemstones identical chemically and structurally with their natural counterparts, came in 1891, but few knew of it. Auguste Verneuil, a French chemist, and his partner E. Fremy had been working on the synthesis of ruby. Both had profited by previous work done by two other Frenchmen, C. Feil and M. Gaudin. This is the way technological discoveries often take place, with each man adding his part to the store of knowledge until at last someone can put all of it to use. In 1891 Verneuil filed sealed papers with the Academy of Sciences in Paris giving full details for his method of making ruby. The method is now known as the flame-fusion process and, with minor alterations, the technique is still followed as he outlined it. It is still the most common technique for growing synthetic gemstones, having been applied to several kinds other than ruby. Verneuil published the contents of his secret papers in November, 1902, and the synthetic-gem industry was launched.

What he had developed was an ingenious chalumeau—an upside-down blowpipe. The technique calls for powdered aluminum oxide, containing a chromium coloring agent, to be sieved down through the flame of the vertical blowtorch furnace. As it passes through the flame at about 2200 degrees centigrade it melts and deposits on a holder to form a single crystal boule, or tapered cylinder, of the synthetic gem material. The word boule is French and means ball—an allusion to the starting shape of the synthetic crystal mass. In a few hours a boule of several hundred carats can be formed. To achieve the greatest degree of perfection in the finished product, very pure materials must be used. Constant feeding of new material to the boule is critical, since even interruptions of fractions of a second can cause color banding in the crystal. The flow of oxygen and hydrogen gas to the flame must be constant, as well as the rate of lowering the holder-pedestal as the boule forms. Variations in all these growing conditions can cause growth stresses in the boule and may suddenly crack it. Through the use of mechanical controls and more reliable kinds of heat sources, it is now possible to make boules in the shape of thick disks up to five inches in diameter, or rods several inches long, for purposes other than gem cutting.

Of course, if the technique works for ruby it should work for sapphire since they differ only in the coloring agent. It does work. Without coloring material a colorless sapphire is produced. Chromium oxide as a coloring agent makes pink sapphires, or ruby if in higher concentration, or green sapphire with an even larger amount. Manganese also makes the sap-

Well-formed hexagonal crystals
of synthetic ruby under various magnifications.
(One at left is enlarged 50 times.)
These were grown by French
chemist-mineralogist E. Fremy and
pictured by him in 1891.

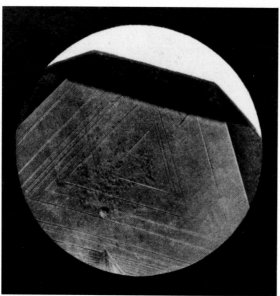 

phire pink, nickel yellow, and titanium plus iron the popular rich blue. All of the natural colors for sapphire can be duplicated, using appropriate coloring agents, as well as several colors not known in nature. Almost every tourist traveling to the orient is tempted to buy a gem of the fabled alexandrite. Few realize they are buying synthetic sapphire to which a trace of vanadium is added. Its color change from reddish under incandescent light to greenish in daylight mimics genuine alexandrite quite well. Actually, whatever color, the bulk of synthetic ruby and sapphire production does not end up as gems but as very hard bearing surfaces, or "jewels," in watches and all sorts of other precision instruments.

For gem purposes, one of the next logical attempts with the Verneuil technique was to see if it could produce star rubies and sapphires. Burdick and Glenn at the Linde Company in California produced the first star gem in January, 1947. Commercial production began later that year. As described in Chapter 2, the star effect in rubies and sapphires is owing to the presence of tiny rutile needles arranged parallel to the three directions of the crystal structure. Man followed nature's lead and introduced rutile—titanium dioxide—into the Verneuil mixture. At boule-formation temperature the rutile dissolves in the melt and, after cooling, the star boules look the same as the others. However, these boules are annealed by heating at about 1100 to 1500 degrees centigrade. At this temperature the rutile separates itself slowly from the rest and forms multitudes of tiny needles properly aligned. Thus, star gemstones are born.

While corundum in its varieties of ruby and sapphire was being synthesized, experimentation was also conducted to see if the Verneuil method could produce other gem materials. It could. The best of these others is

synthetic spinel. Both Verneuil and his student L. Paris had worked on producing good blue sapphires. In the process, synthetic spinel was made accidentally. Until 1930, however, the discovery languished and even then commercial production was used mostly for imitating other gems. Spinel is a magnesium aluminum oxide with one magnesium atom to every two aluminum atoms in its composition. It was found that this proportion does not produce good boules in the Verneuil furnace, and normally one magnesium atom to five aluminum atoms is used instead. The extra aluminum stresses the spinel structure enough so that its density and refractive index can be distinguished from those of natural spinel. Flame-fusion spinel, like corundum, is made in many colors. It successfully imitates the aquamarine, blue zircon, green tourmaline, chrysoberyl, and others. Colorless spinel can be confused with white sapphire and sometimes even with

diamond. Since the 1950's a good red spinel, close to ruby in color, has been made and used as a substitute for ruby or for natural red spinel, which is the most popular color.

Two other synthetic gems have been produced by the Verneuil technique. Neither of these has a natural counterpart but both have found their way into the gem trade. They are titanium dioxide and strontium titanate. Titanium dioxide has an exact match in nature called rutile, except that natural rutile tends to be such a very dark reddish brown as to verge on black, and is quite opaque. Synthetic rutile, on the other hand, is transparent and almost colorless. As the boules come from the furnace they are opaque and black because of a deficiency of oxygen in the rutile structure. By reheating in a stream of oxygen, the boules begin to change color from black to dark blue to light blue to green, pale green, and finally yellow, and almost colorless. Occasionally reddish or orange boules result. If the heating is stopped at any point in the color change sequence, gemstones of the selected color are made. Synthetic rutile has one of the major disadvantages of natural rutile for gem use. It is too soft. However, it has extraordinary color dispersion—approximately six times as much diamond—which endows its gems with great fire. Synthetic rutile is marketed as Titania. It is not usually sold as a substitute because it is too easily distinguished from all other gems by this unusually strong fire.

The other Verneuil synthetic, strontium titanate, is sold under various names of which Fabulite is perhaps best known. Several of the names attempt to suggest that it is some kind of diamond. It is not related at all to diamond and actually has no counterpart in nature. Its value as a diamond substitute is obvious since it has a dispersion four times that of diamond with resulting high fire, and is completely colorless like the best diamonds. Titania, its only rival, has far too much fire and is never completely colorless. Both Fabulite and Titania can never be more than substitutes because both have a hardness of only 6.

Other Gem-making Processes

Two other crystal-growing techniques have been developed which have successfully produced synthetic gems. They are the flux-fusion method and the hydrothermal method. A close look at the names of the methods gives an indication of what they are. Flux-fusion growth is carried out in melted solvents—"flux" being a word for any substance that promotes melting, and "fusion" meaning melting by heating. Hydrothermal growth employs hot water solutions usually under pressure—"hydro" meaning water and "thermal" meaning heat. Many minerals do not dissolve well in water. To overcome this difficulty, other dissolving agents or fluxes are used at high temperatures, as in the technique developed by the Bell Laboratories for growing synthetic ruby. A platinum container about 10 inches high and 6 inches in diameter is filled with a flux of eleven pounds of lead oxide (PbO) and fourteen ounces of boron oxide (B_2O_3). To this is added the material for making the ruby itself: twenty-six ounces of aluminum oxide (Al_2O_3) and just a trace of chromium oxide (Cr_2O_3) to give a red color. The container and

its charge are then heated in a electric furnace to 2400 degrees Fahrenheit, so that the chromium oxide and aluminum oxide dissolve in the now-liquid flux. During six hours of heating at this temperature the contents are stirred continuously by rotating the pot first one way and then the other. After six hours, the temperature is slowly and evenly dropped about 170 degrees Fahrenheit per day for several days. As the temperature drops just below 2260 degrees Fahrenheit, ruby crystals begin to form. When cooling is finished and the whole mass has become solid, with the new ruby crystals mixed in it, the flux is dissolved by soaking in nitric acid, which has no effect on the ruby crystals.

The story of the development of synthetic emeralds began with a small-scale flux-fusion process but made a dramatic advance when investigation apparently turned to hydrothermal methods. Paul Hautefeuille and Alexis Perry of France are credited with producing the first true emeralds in 1888. Their tiny crystals were made by heating appropriate raw materials for fifteen days at a temperature of 1480 degrees Fahrenheit. The materials included not only those needed for the beryl but also chromic oxide to give it emerald color and lithium molybdate to act as a high-melting-temperature solvent or flux. At the end, everything but the emerald was removed by treatment with hydrochloric acid. The emeralds were excellent but much too small for gem purposes. Other chemists continued to work on emerald production until finally chemists of the I. G. Farbenindustrie in Germany made excellent, cutting-quality emeralds in 1934. The com-

78

pany never divulged the secrets of its process and soon seemed to lose interest in it. A very few commercial gems were made and marketed under the name "Igmerald." These are rare and have now become collector's items. At almost exactly the same time, Carroll Chatham, a San Francisco chemist, produced some fine gem emeralds. He is still producing them but has increased the size and quality of the stones through the years. His process, too, is secret, although he says that it is hydrothermal. It could very well be hydrothermal and, if so, parallels can be drawn between this process and the known methods for growing gem-quality quartz crystals.

In growing quartz crystals very high temperatures and pressures are needed to get the raw materials into solution in water. Large quantities of valuable quartz crystals are grown this way for use in the electronics industry. Equivalent natural quartz of the type needed is so rare and so expensive that it is economically sounder to manufacture it. Pressures up to fifty thousand pounds per square inch are used at temperatures of as much as 1300 degrees Fahrenheit, and substances are present to help the quartz go into solution. Under these extreme conditions the auto-claves, or "bombs," containing the solution must be extremely strong. They must have leak-proof and blowout-proof closure devices and must be silver- or platinum-lined to prevent contamination of the solution by chemical attack on the lining by the mineralizer. Usually about 6 inches in diameter and twelve feet long, industrial bombs have 4-inch-thick walls that are made of cold-rolled steel. The bomb is loaded with water, mineralizer, and a quantity of small pieces of pure natural quartz in the bottom. At the upper part of the bomb are hung the seed crystals of quartz cut in the form of flat plates. The bomb is sealed and heated electrically to crystallization temperature. The pressure automatically rises tremendously because of the heat-expanded contents. After about twenty days the cycle ends. All the small pieces of natural quartz from the bottom have been dissolved in the water solution and re-deposited on the surface of the seed crystals to make fat new crystals of quartz.

An interesting addition to techniques for growing emeralds for the gem trade was developed around 1960 in Austria. It, too, is probably a hydrothermal method but differs in using cut beryl gems of poor color as seed material. For a few days the cut stones are kept in the hydrothermal solutions and grow a thin coat of emerald. When the gems are removed, a final polish is all that is needed to prepare them for market. They are beryl all the way through and the center part is even natural beryl. They have all the necessary characteristics of emerald—color, refractive index, hardness, specific gravity, composition—and in addition may even show natural inclusions in the body of the gem.

Success is already assured for the commercial production of ruby, sapphire, spinel, and emerald. Other synthetic gems are in an active stage of development with a high probability of future production. Diamonds are being grown in quantity for industrial purposes. At the moment, almost all the crystals are too small for gem purposes. As equipment and methods improve, they should be made in larger sizes. Even then, however, it is questionable whether the stones will compete with natural diamonds for gem use. Of course, this assumes that the natural supply will continue to be available.

Small crystals of alexandrite and other chrysoberyl, zircon, and garnet have already been grown from fusions. After a period of time it is reasonable to think that improvement in techniques will add all of these to the list of commercial synthetic gems. Garnets have just recently reached this state of development with production of gem-quality YAG (yttrium aluminum garnet).

Treated and Altered Gems

Heating reddish-brown zircon pebbles in a crude oven in a small oriental village seems a far cry from subjecting an off-colored diamond to bombardment in an atomic pile in the United States. These operations are alike in their purpose—to alter the natural color of the gem so that it will be more desirable. All of the beautiful blue and colorless zircons we know in the gem trade are so attractive because they have been heated. Control of the heating conditions in the crude furnaces will determine whether they are to be blue or colorless. The diamond bombardment, although it is somewhat predictable in its results, is a bit more of a gamble. If it does succeed in converting a poorly colored gem into a rich yellow or green the results are both economically and aesthetically pleasant.

When the color tampering is not too obvious it seems acceptable in the gem trade. For example, no one objects that the bulk of the finest colored aquamarines in a jewelry shop are stones that have been heat-treated to intensify their pale color or to make them more blue than green. The new gem Tanzanite often gets its rich sapphire-blue color from heat-treating. Pale, unevenly colored amethyst can become golden brown citrine the same way.

Less acceptable coloring is done by dyeing. Most hard, faceted gems will not absorb dyes but there are a number of other, more porous gemstones. Turquoise takes on a deeper color when dyed, or it may even be helped by soaking in melted paraffin wax or plastic. Gray or pale green jade is easily dyed green to achieve the strong color of the best jade. This is especially true of the more opaque varieties of jade. Quartzite can also be dyed jade green with ease and substituted for actual jade. Agate in all its many varieties can be dyed almost any desired color. Blue agate—a very unnatural looking color for agate—is made by soaking the stone first in a solution of potassium ferrocyanide followed by soaking in a solution of ferrous sulfate. Red is made by soaking the agate in ferrous sulfate and heating to a high temperature. Even black agate, looking like natural black onyx, can be made by boiling it first in a sugar solution and then soaking it in concentrated sulfuric acid. The acid converts the soaked-in sugar to black carbon.

When color cannot be introduced by dyeing it is sometimes painted on. It is a common trick to touch the back facets of a yellowish diamond with violet dye to make the entire stone seem colorless. A black back painted on a clear Mexican opal may help its weak color display to look better. Worst of all, the color in a gem may be introduced by backing it with a piece of foil in the setting. In the former crown jewels of Russia is the famous, ten-carat Paul I Diamond, once reputed to be a fine example of a ruby-red diamond. After the Russian revolution, during inventory, it was found to be a pale pink diamond backed with red foil. Even royalty apparently had trouble with gem counterfeiters.

From the discussion in this chapter it is obvious that the problem of gem substitutes is really several problems. Certainly, synthetic gems are beautiful and desirable objects in their own right. They are certainly not in the same class with dyed or foiled or glass or plastic gems. Perhaps they are no more undesirable than a natural gem that has been altered in color by heating. Slowly but surely the gem-buying public makes up its mind about these things, deciding what is and what is not acceptable substitution. To guard against fraud in the process there are organizations, both Federal and private, to assist the trade and the public. The most important of these in the United States for testing gemstones, detecting counterfeits, and generally keeping track of new developments in gem substitution are the Gem Trade Laboratories of the Gemological Institute of America in New York and Los Angeles. Their efforts and those of many others help to support public confidence that there are differences between natural and substitute gems and that the experts are able to recognize them.

The Elite Minerals|4

There is a rather common idea that new discoveries of gemstones are made accidentally after long and diligent search over all sorts of outcroppings of rock on the earth's surface. Although this impression is naïve, it is not too far from the truth. Often, gem deposits are found accidentally. The colorful story of the accidental discovery of gorgeous, flashing, Australian opal in 1875 by an itinerant miner named Paddy Green is typical. In most cases, however, the gem prospectors who make the discoveries could have saved themselves considerable time, work, and expense, and increased tremendously their chances of finding something of value. It only takes a little knowledge about the way gemstones form in the earth's crust to be able to anticipate the kinds of rock deposits in which they are likely to occur. Gemstone-bearing rock deposits are mineralogically unusual and are therefore rarely found. Those rock deposits which cannot possibly contain gems can be passed over by the prospector, and potential gemstone-producing deposits can be explored in more detail, thus greatly increasing the possibility of discovery.

The geologist often prefers to classify rock deposits according to the three major processes by which they are formed—igneous, sedimentary, and metamorphic. Each of these processes, however, may involve variations, any one of which, even if slight, may be sufficient to guarantee that gems will not be found in a particular deposit. Petrology, the study of rocks, has determined and described large numbers of different kinds of rocks, all of them formed by one of the three basic processes. Most of these rocks are of no significance in a discussion of gemstone occurrences. A sampling of those that are is given here.

Igneous Rocks

When our earth was young it was a ball of hot, melted material consisting of one great soup of elements. Cooling, by loss of heat into space, eventually began to have an effect, and the first, cooled, solid rocks were formed. Rocks like these, formed from molten material, are called igneous rocks. The word igneous comes from Latin and means fiery. Later, as the earth aged, other geologic processes took over to help shape and reshape the ever-changing crust on which we live, and to attack and alter the original, exposed, igneous rocks. However, igneous processes continue even now, billions of years later. The earth has a vast store of molten material lying not too far beneath its thin, solid crust that varies in thickness from roughly ten to thirty miles. Now and then a quantity of this molten magma begins to move upward through the crust toward the surface. The reasons for this upward movement are not yet clear. Sometimes the material never quite reaches the surface but pauses, cooling slowly to form an enormous solid mass of igneous rock called a batholith. If weathering and erosion subsequently strip off the crustal covering, the batholith may be exposed. The great exposed Idaho batholith, for example, makes up about 16,000 square miles of the Salmon River mountain country. The enormous Confederate Monument in Georgia is carved into the face of Stone Mountain, an exposed granite batholith 630 feet high.

Often, as the magma pool moves up through the crust it finds weak spots. It sends off smaller masses and injections of molten material under high pressure. These, too, may never reach the surface. They cool gradually and form other, smaller deposits of igneous rock. All of these deposits, which remain embedded in the crust or are exposed only by erosion, are called intrusive rocks. But now and then magma reaches the surface and bursts forth into the air in a magnificent volcanic display of sound and fury, pouring out untold tonnages of ash, steam, gases, and lava. The volcano forms extrusive igneous rocks by dumping its rapidly cooling contents onto the surface. These outpourings can be incredibly large. The Columbia Plateau in the northwestern United States was created by a long series of lava floods which piled basalt layers more than five thousand feet thick over two hundred thousand square miles.

In general, the extrusive igneous rocks are poor hunting grounds for gemstones. Because of their sudden exposure and rapid cooling, the individual minerals that make up the rock mass have not had time to grow into large enough crystal grains to be usable as gemstones. Often the cooling is so rapid that the rock is glassy and individual mineral grains are not formed at all.

It is among the intrusive igneous rocks, brought near the surface or exposed to the prospector by erosion, that gemstones may more likely be found. Obviously, all these rocks will not be gemstone producers because the appropriate elements might have been missing from the original batch of magma. The most productive kind of intrusive igneous rock is called a pegmatite, and it too starts its existence as magma. As a magma begins to cool, it forms the solid crystals of minerals. Quartz, feldspar, and mica very commonly form at this stage, often producing a rock known as granite. The liquid part remaining as the granite forms becomes richer and richer in less common elements, and in water, carbon dioxide, other gases, and certain acids. This residual hot liquid is forced out of the solidifying mass and injected under great pressure and at high temperature into the surrounding crust. Eventually this material cools and crystallizes separately in a unique way.

These last-formed minerals tend to be in very coarse crystals, many of them of unusual size and composition and some of them gemstones. Such coarsely crystalline deposits are called pegmatites and they may range in size up to bodies that are thousands of feet long and hundreds of feet thick. Quartz, feldspar, and mica make up the great bulk of pegmatites. A barren pegmatite may contain only these three, but a "dirty" pegmatite, the dream of the gemstone miner, may contain tourmaline, topaz, beryl, or perhaps even garnet, spodumene, iolite, orthoclase, or chrysoberyl. The famous gemstone-bearing pegmatites of southern California and Brazil are typical of this kind of deposit.

Sedimentary Rocks

Igneous activity seethes in the crust and startles us by disrupting the tranquility of the surface. In reality, however, the surface is far

85

from peaceful. Every rock, the instant it is exposed to air and water and weather, is subject to constant attack. Few rocks can stand up under the assault. Slowly fragmented, broken off, torn, ground, and worn, all the crustal rocks are reduced to rubble. Carried off by wind, water, and gravity, the rubble, varying in size from mud particles to large boulders, is finally deposited somewhere. Other material, carried off dissolved in water, is later deposited to add its bulk to the rubble, or even to cement muds, sands, or gravels together. Eventually most of the deposits form a new series of sedimentary rocks.

Most of the earth's sediments are worthless as gemstone sources. But some few, such as the gravel deposits of Burma and Ceylon, have been mined for centuries and are still great treasure troves of gemstones. The gem productivity of any sedimentary deposit depends entirely on the original rocks from which they came. In Ceylon for hundreds of millions of years erosion has been tearing away the highland mountains. The sediments have built thick layers in the V-shaped river valleys and as much as twenty feet of overburden in the southern bottom lands. The rich red muds of the land around Ratnapura are covered by rice paddies, and under the paddies lie the gemstone gravels. There is a wide variety of gemstones found there, including sapphire, ruby, chrysoberyl, spinel, zircon, topaz, garnet, tourmaline, quartz, moonstone, sphene, iolite, fibrolite, andalusite, diopside, kornerupine, apatite, and sinhalite. These are the most extraordinary gem-bearing sediments in the world. Certain sediments along the coast of

Africa are probably more famous. These are the river gravels—hunting grounds of the diamond prospectors for many years—from Hopetown, along the Orange River to its junction with the Vaal, as far as the town of Pontcherstroom, that early diamond prospectors combed back and forth. In these locations diamonds occur because of millions of years of erosion which have torn away thick layers of the surface of Africa.

Metamorphic Rocks

If the earth's crust were motionless and unchanging there would be only the original igneous rocks exposed at the surface. Erosion banishes any hopes for crustal stability by moving incredible tonnages of material from here to there through millions of years. To compensate for all of this shifting weight, the thin crust, floating on its yielding foundations, bends, shifts, buckles, cracks, and wrinkles constantly—sometimes with great shaking and rumbling and growth of mountain ranges. Of course, the rocks of the crust are subjected to terrible pressure and temperature changes during these events. The forces are applied slowly but powerfully. The rocks react by shuffling their elements into new combinations able to exist under the new conditions.

The intrusion of a hot, igneous mass into pre-existing rock can have the same powerful effect as crustal movement, and in addition may bring in new materials to add to the rock mix. Even sediments, piling up for hundreds or even thousands of feet, bear down with such great weight on the sediments beneath that they too are stressed beyond endurance and must change. Imagining the great pressure under a thousand feet of water, consider what it is under a thousand feet of rock. In deep mines, because of this pressure, there is always a danger of "rock bursts." These are explosions of rock from the roof or walls of a tunnel caused by the great pressures above.

All such pressure- and temperature-transformed rocks are called metamorphic rocks. Sometimes they do bear gemstones. Ruby and sapphire in Burma were formed in a metamorphosed limestone from which they erode and are concentrated in stream gravels.

For all the earth's rock-making processes, gemstone-quality mineral specimens are very rare. Of the less than two thousand mineral species known, only a handful—one hundred at most—have been found suitable for gem purposes. Of this group, only fifteen or so are normally found circulating widely in the gem trade. Some forty more are sometimes available to the collector and connoisseur. The remainder are more correctly considered decorative or ornamental stones of relatively low intrinsic value. Of the select hundred, only very small amounts are ever found of adequate quality for gem use.

The gemstones are reviewed in this and the following chapter in the order of importance seemingly assigned to gems by those who acquire them, that is, their commercial, collectible, and decorative interest.

The Commercial Gems

Amber: The passage of millions of years, perhaps twenty million or more, has robbed the persistent, resinous remains of the sap of certain plants—such as *Pinus succinifera,* the amber pine—of most of its liquids and gases. What is left is a solid mixture of an organic acid called succinic acid, a number of resins and oils. The mixture, called succinite amber, has a hardness slightly over 2. This is much too soft for most conventional jewelry uses, but the material is often made into beads and pins in which it gets support and protection from the metal mounting. Amber is usually yellow to brown in color and varies from clear and transparent to opaque with an oily luster. It softens at a temperature well below that of boiling water. This low melting temperature makes it possible to fuse and press together small fragments of amber into larger, more usable masses. In Germany amber is called "bernstein" because it burns easily, emitting a pleasant, aromatic odor. Often amber pieces will have in them the fossil remains of insects that were attracted to the sticky, aromatic sap and caught in it to be preserved in all their delicacy for eons.

Copal resin, also exuded from certain trees, is much like amber and is the most common natural substitute for it. The best copal, like succinite, is fossilized but is very new as compared with the ancient succinite ambers. Currently, Bakelite and several other plastics can be manufactured to resemble amber very closely. It can be appropriately colored and stressed, and may even have insects added—modern insects, of course. The aromatic smell produced by burning a small sample of copal or amber separates them rather easily from the plastic imitations.

When the Teutonic Order of Knights

conquered Prussia in the thirteenth century its most important acquisition was a monopoly on the highly prized Baltic amber supply, which for centuries had yielded enormous financial returns to those who controlled it. Now part of Russia, the Samland coast of East Prussia is the principal source, although it does occur elsewhere. Burmese amber, or "burmite," from the Hukong Valley, is redder than Baltic amber. It is also harder and denser. "Simetite," a reddish brown Sicilian amber, is found along the Simeto River. Minor amber deposits are scattered from Europe to Maryland, New Jersey, and New York, to Cedar Lake in Manitoba, Canada, and on around the world in many other unheralded places.

Beryl: It seems odd that a gemstone so important in commerce as beryl should seldom be advertised under its true name. Jewelers and potential customers use the special names given to its color varieties. The best known of these are emerald (green), aquamarine (green to blue green), and morganite (pink). Names for colorless beryl (goshenite) and golden beryl (heliodor) are gradually being retired.

All beryl, a beryllium aluminum silicate, has a hardness of 8 and no tendency to cleave or to be particularly brittle. This makes it one of the more durable gemstones. The specific gravity is considerably less than three-fourths that of ruby and sapphire, so that stones of larger size can be worn in jewelry with comfort. Since it occurs in large crystals of excellent quality its gem potential is high. Also there is enough of it in the gem mines of the world to supply the expanding market.

The great gem merchant, J. B. Tavernier, who traveled in the Far East in the middle 1600's, was well aware of the source of most of the important emeralds—if not the route of their transport. He wrote that "as for Emeraulds, it is a vulgar error to say they come originally from the East. And therefore when jewellers and gold-smiths, to prefer a deep-coloured Emerauld enclining to black, tell ye, it is an oriental emerauld, they speak that which is not true. I confess I could never discover in what part of our continent those stones are found. But sure I am, that the Eastern part of the world never produced any of those stones, neither in the Continent, nor in the islands. True it is that since the discovery of America some of those stones have been often brought rough from Peru to the Philippine Islands, whence they have been imported into Europe; but this is not enough to make them Oriental."

Crystal specimens of emerald acquired by Cortez came to Europe by a more direct route and are still carefully guarded at the Natural History Museum in Vienna. Tavernier himself carried a large number of fine emeralds into the Orient. Somehow, Tavernier did not seem to know of the ancient emerald production of Cleopatra's mines in North Africa, although he may have discounted it because the quality of the emeralds obtained was not too high. Mas'ûdi, an Arab author of the tenth century A.D., tells of an odd belief that these Mount Zabarah emeralds were found in quantities depending on the seasons, prevailing wind direction, and other atmospheric conditions, and that the intensity of their color changed with phases of the moon.

*Opposite: 14-inch gem-cutting-quality
crystal of aquamarine. Below left:
31-carat orange, or "padparadscha," sapphire.
Right: Huge, 2054-carat, step-cut beryl.
Bottom: Step-cut, 911-carat aquamarine.
All are from Brazil except Sinhalese sapphire.*

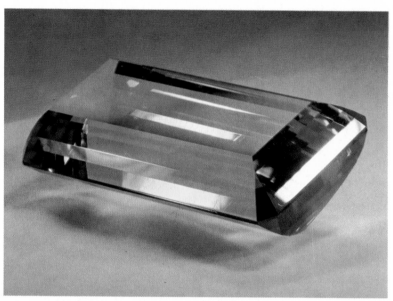

90

Russian emeralds were found much later, in 1830, along the Takovaya River, northeast of Sverdlovsk. The crystals were large but of low quality, so that only small stones could be cut from them. They are still found there in a rather coarse metamorphic rock called mica schist and associated with chrysoberyl and common, nongem beryl. At the end of the 1800's and beginning of the 1900's more emerald was found occurring in pegmatite veins near Emmaville, New South Wales, and Poona, in Western Australia. The life of these mines was short and their production minor. African production of emeralds proved a bit better. In 1927 the Lataba District of northeast Transvaal began to supply emeralds from mica schists near pegmatite contacts. Production was sporadic and quality pieces were too few to make the mines profitable. Sandawana emeralds from Southern Rhodesia, found in 1956 in schists near a pegmatite, are of fine color and are still being marketed in small, quarter-carat sizes. There is some doubt about whether the quantities of emeralds used in India for hundreds of years ever were mined there. In recent times, however, emeralds—some of good quality and large size—have been found in Udaipur and Ajmere-Merwara in bands of mica schist.

By and large, a survey of world production of emeralds is not encouraging except for Colombia, the undisputed, all-time champion of emerald producers. These emeralds and their sources are covered in detail in Chapter 6.

The green color of emerald is due to the presence of a very small amount of chromium in its composition. Very likely an even smaller amount of iron impurity helps to determine the stone's ultimate shade of green. On the other hand, if chromium is absent, iron alone can cause the greenish-blue tinting of this mineral. If so, the material is called aquamarine.

Unlike emeralds, aquamarines are fairly common. On the island of Madagascar alone there are several dozen sources. The state of Minas Gerais in Brazil sometimes seems to produce an unending supply. A list of aquamarine discoveries in the United States almost seems a roll call of the states. Usually, aquamarine is found in pegmatites rather than in the schists so familiar as a source of emerald. The most popular aquamarine color seems to be a soft blue, sometimes tinted slightly green. Almost every good blue aquamarine seen in commerce, however, gets its color from heat-treating—between 400 and 450 degrees centigrade—of certain greenish or off-yellow stones. The treatment changes chemically and permanently the nature of the iron impurity and produces a good bluish color. Synthetic spinel can be manufactured with the same color, making it a very deceptive substitute.

Morganite, named after J. P. Morgan, the banker and gem connoisseur, gets its peach, rose, or pink color from traces of lithium oxide in the beryl composition. Its density, its refractive indices, and so forth vary only slightly from those of other beryls. It is rarer than aquamarine, but does occur more plentifully and in larger fine-quality stones than emerald. The pegmatites of Brazil, Madagascar, and California supply most of it.

Chrysoberyl: Because it gives us three distinct and attractive varieties, chrysoberyl is also noteworthy among gems. Ordinary gem chrysoberyl occurs in various shades of green, brown, and yellow, and is attractive enough, except that its two other unusual varieties—cat's-eye and alexandrite—overshadow it. The mineral is a beryllium aluminum oxide related chemically to the gemstone spinel. With a hardness of 8½ it is third hardest of all the gems after diamond and corundum. This hardness, combined with no particular tendency to cleave and its strong resistance to chemical attack, makes it a remarkably durable gemstone.

Chrysoberyl cat's-eye gems command relatively high prices almost everywhere around the globe because of their rarity and beauty. Known in antiquity as *oculus solis*—"eye of the sun"—the gem's sharp ray of light superimposed on a green to yellow-green to brown cabochon background produces a spectacular effect. Apatite, diopside, quartz, tourmaline, scapolite, moonstone, and occasionally even beryl and other gemstones exhibit a cat's-eye effect because of light reflection from a myriad of needle-like inclusions oriented in a single direction. (This phenomenon, called "chatoyancy," is explained in greater detail in Chapter 2.) The chrysoberyl cat's-eye is most highly prized, and the term "cat's-eye" in the gem trade is reserved solely for this species. Ceylon gem gravels have been the most important source of cat's-eyes since earliest times. Even as late as the early 1800's a prime jewel in the crown of the kings of Kandy, in Ceylon, was an enormous cat's-eye.

Alexandrite is a darkish colored chrysoberyl with the remarkable characteristic of changing its color from red in artificial light to green in daylight. The delicate red-green color balance—emphasized one way or the other when viewed in a light source rich in blue and green or one rich in red—is caused by the presence of minute amounts of chromic oxide. Although the Russian czars are no longer with us, the name of Alexander II is immortalized because he came of age the day the gem was discovered in 1830 at the emerald mines near Takovaya in the Ural Mountains. Although first found in Russia, much larger alexandrites have since been found in the gem gravels of Ceylon. The largest of these on record is a beautiful 66-carat cut stone in the gem collection of the Smithsonian Institution. The less spectacular color change of the Ceylon stones makes them less desirable, although they are still very rare and expensive. Undoubtedly, the large majority of cut alexandrite in commerce today is an amethystine-colored synthetic sapphire made in an attempted simulation of the real gem.

In addition to the Ural Mountains of Russia and the gem gravels of Ceylon, gem varieties of chrysoberyl are found at Mogok in Burma and especially in the state of Minas Gerais and elsewhere in Brazil. Minor occurrences have come to light in the United States, Madagascar (Malagasy Republic), Zambia, Rhodesia, and elsewhere.

Coral: Somehow coral seems less attractive as a gem material when we learn that the ancient Greeks thought it was blood which had become hardened after dripping from the severed head of Medusa. The fanciful idea does, at least, suggest an organic, living origin for coral if nothing else. Although it is used for

94

Opposite: 114-carat chrysoberyl from Brazil. Bottom left: Rare, 4-carat demantoid garnet from Russia. Right: 42-carat sapphire from Ceylon. Below: Fine chrysoberyl cat's-eye from Ceylon. Bottom: 138-carat Rosser Reeves Star Ruby.

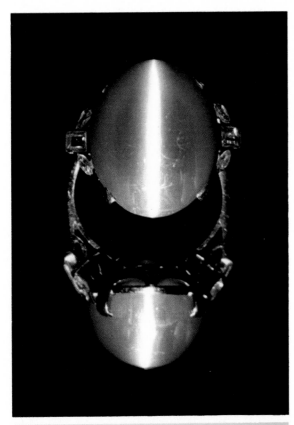

gem purposes, coral is not technically a mineral, since it is manufactured by small sea animals. Large colonies of these animals secrete a hard skeleton of calcium carbonate which they deposit for support as twiggy branches, fans, and other forms, depending on the particular species of animal involved. Most precious for gem use are the bright red, branched deposits of the *Corallium rubrum,* or red coral animal. Whether it be "rosso scuro" (dark red), "rosa vivo" (bright rose), or "rosa pallido" (pale rose), all of this coral has a hardness of about 3½. The coral animal thrives in its colonies only in relatively quiet and clear ocean water which holds a temperature between 55 and 60 degrees Fahrenheit. Malaysian, Japanese, and Mediterranean waters supply most of the marketable coral, with the very finest being dredged off the coast of Algiers and Tunisia and several points along the Italian, French, and Spanish coasts. Italy has long maintained a near monopoly in the processing and carving of Mediterranean coral.

There are other corals than red used for gem material, including a white variety deposited by *Oculinacea vaseuclosa,* a black variety by *Antipathes spiralis,* and blue by *Allopara subirolcea.* Black and blue varieties, including the black Hawaiian variety, are not composed of calcium carbonate like the choice red varieties, but appear to be made of a darkish organic material.

Corundum: According to the ancient Persian view of the Earth and its surroundings, the great globe sits on an enormous blue sapphire which gives color to the sky itself. It has always seemed the fashion to associate the sapphire with the shades of blue. Even the name "sapphire," translated from Latin, means blue, and in the gem trade sapphire is generally assumed to be blue. Technically, as mentioned earlier, it can be any color except red. When red, it is called ruby. Ruby and sapphire are merely gem varieties of the same mineral called corundum.

Even from before the time of written history ruby and sapphire have been treasured. In the Mogok district of Upper Burma, source of an incredible treasure of ruby and sapphire, artifacts scattered in the mining area date activity far back into the Bronze and Stone Ages. Both Oriental and Western literature abounds in references to the ruby and sapphire. Royal regalia through the ages—from the crowns of Lombardy and Hungary and the Gothic kings to the present-day crowns of England and Denmark—has carried these gems.

There are good reasons for the continued high popularity and great value of ruby and sapphire through all of man's recorded history. Corundum, a natural aluminum oxide, is hard. Next to diamond it is the hardest natural substance known. The mineral is also rather tough, with little tendency to crack, chip, or cleave. Obviously, it is a very durable substance that can stand the roughest abuse.

Well-colored corundum is as beautiful as any tinted gemstone can be. Bright red and bright blue draw the highest praise, but the softer shades of blue and violet, pink and lavender, rose and yellow, purple and green offer a very pleasing choice of colored gems with identical durability. There is even a gorgeous orange variety called "padparadscha"

from the Sinhalese name for the lotus blossom. All of these colors arise from the presence of very small and varying amounts of titanium, iron, and chromium compounds in the corundum. Traces of chromium oxide give ruby its popular color.

The beauty of ruby and sapphire is sometimes enhanced by asterism—a bright star-like reflection—as described in Chapter 2. Although a star may be found in almost any color of corundum it seems to show best against a strong red or blue background.

Durability and beauty alone could account for the esteem in which gem corundum is held. However, it is also rare. Corundum itself is a very common mineral occurring round the globe in many deposits. In Macon County, North Carolina, for example, an enormous deposit of the mineral was found in 1858 by Colonel C. W. Jenks. Tons of corundum were removed from these deposits, including some very large crystals. One of these is recorded as weighing 312 pounds. And yet in all of North Carolina no satisfactory gem sapphires have ever been found and only a few, very small, fair-quality rubies.

The finest rubies, and some feel the only fine rubies, are found in the gem gravels near Mogok, Burma, where they were deposited after eroding out of metamorphic limestones. The pigeon's-blood ruby from these deposits has a strong red color tending slightly toward purple. Such ruby can command prices comparable to those of the best diamonds because there is so little of it. Thailand rubies, coming from the area around Chantabun, are somewhat more brownish. The best Ceylon rubies, from the gravels of Ratnapura and other areas in the southern part of the island, where they lie weathered out of metamorphic rocks, are a strong rose-red color.

Gem sapphire occurrences roughly parallel those for ruby. In Burma, sapphire seems more concentrated in deposits a few miles away from Mogok and at a higher elevation. Thailand, next most important producer after Burma, has yielded very fine sapphires from beds of coarse sand covering a large area centering around Battambang and spreading over into Cambodia. Superb sapphires of the best "Kashmir blue" color are recovered from the northwest Himalaya mountains in Kashmir. There they are found in a pegmatite with other gem materials, such as tourmaline and garnet. Ceylon, too, produces good sapphires in many colors with pale blue, pink, yellow, and white being the most common. Other occurrences should be mentioned, including the finest green sapphires—from the gravel fields of Australia around Anakie and elsewhere. Small but good gem sapphires with a peculiar bright luster come from weathered igneous rocks at Yogo Gulch in Montana. In recent years gemmy crystals of both ruby and sapphire have been found in Tanzania. Every occurrence, no matter how important, yields only a very small number of suitable gems from the large tonnages of rock that must be processed. Corundum gems are rare, indeed!

Diamond: The value and rarity of diamond have always been legendary. To read a book such as *Diamonds—Famous, Notable and Unique,* by Lawrence Copeland, is to read a fascinating account of important people,

places, and events almost from the beginning of recorded history. This fabulous substance was introduced to the Western world from India. Written evidence at the source indicates that diamonds were mined there at least as far back as 400 B.C. Through the years, Europe had heard rumors of Indian diamonds, but there was no confirmation until J. B. Tavernier first wrote an eyewitness acount of them for Europeans about the middle 1600's.

Diamond is a variety of the element carbon with its atomic structure assembled in such a way as to produce transparent crystals of the hardest substance known. Hard as it is, this mineral is very resistant, but not totally impervious to damage. It is brittle and has a cleavage—called "grain" by professional diamond cutters—so that under the severe shock of a sudden, sharp blow it may crack, chip, or even cleave. The tendency to cleave is sometimes used by diamond cutters to split certain pieces of rough diamond to appropriate sizes and shapes for cutting. Fortunately, normal hard wear doesn't faze this durable gem. Carbon in this form has to a high degree everything else a mineral substance requires to be considered a gemstone: brilliance, beauty, rarity, and portability. A high refractive index and high dispersive ability combine to give diamond the beauty of brilliantly colored flashes.

Occasionally, nicely colored yellow, green, blue, or pink stones are found, thus adding to the aesthetic possibilities of the gem. Most commercial gem diamonds are colorless or very pale, steely blue. When they are properly cut, lacking in inclusions, and especially when they are impressively sized, they command prices at the top of the scale for gemstones. Strongly colored stones called "fancies" bring even higher prices. Red, pink, and blue fancies are the rarest of all.

We hear so much about diamonds in the world of gems that they seem more common than they are. They are very rare even in South Africa, which is the prime producer in our day. It has been variously estimated that at least one ton, and in some places as much as twenty tons, of ore must be handled to recover one carat of diamond. Even then, most of this is industrial quality diamond unsuited for cutting gems. The deposits themselves have proved hard to locate and only a half dozen places around the world have ever been found. Flowing through the markets of Golconda, India, for centuries came an estimated twelve million carats of diamond from this earliest source. The recovered treasure included the 109-carat Koh-i-noor Diamond, now in the Queen's Crown of England, the 44½-carat blue Hope Diamond, now in the Smithsonian Institution, and several other famous diamonds. Exhaustion of the deposits and discovery of new ones in other parts of the world have reduced present recovery to a small handful each year.

The diamond-mining center shifted abruptly with the shipping of a quantity of diamonds in 1727 from the gold-prospecting area near Tejuco—now known as Diamantina—in the state of Minas Gerais, Brazil. As in India, diamond mining and recovery were rather primitive processes using manual labor for mining and washing the gem-bearing gravels.

Brazil, too, in its time yielded important diamonds. The best known, and certainly a large one at 726 carats in the rough, is the President Vargas. Twenty-nine gems were cut from it by Harry Winston, Inc., in 1941.

The entire diamond mining picture changed rather quickly again in 1867, when a rush for South Africa began after exhibition of the newly found, 21½-carat Eureka Diamond in Paris. It had come from gravels at the headwaters of the Orange River, picked up by a boy named Erasmus Jacobs. Thereafter, discovery followed discovery and such famous mines as DeBeers, Bultfontein, Jagersfontein, Dutoitspan, Kimberley, and Premier were opened. At first, stones were recovered only from gravel deposits. Later they were found in and separated from a yellow clay. Still later it was discovered that the yellow clay was merely a weathering product of the harder blue-gray igneous rock called "kimberlite," which lay beneath the clays as enormous plugs in ancient volcanoes. The original rock, of which most of the kimberlite consists, is called peridotite and is composed of the mineral species olivine, with some enstatite, chrome diopside, and other substances. Hard-rock mining was then begun in the kimberlite. Nowadays mining has even moved to dredging of sea-bottom gravels deposited along the coast as a result of millions of years of erosion.

The most spectacular diamond ever discovered in South Africa was the Cullinan, which weighed more than one and a third pounds before cutting and whose shape suggested that it was part of an original crystal twice as big. It was found in 1905 at the

Premier Mine and named for Sir Thomas Cullinan, the chairman of the parent Transvaal Company. The government of Transvaal bought the stone and gave it to King Edward VII of England in 1907.

When cut, the Cullinan yielded one hundred and five stones, four of them now among the British crown jewels. The Star of Africa, at 530.2 carats the largest cut diamond in the world, is set in the King's Royal Scepter; a 317.4-carat gem is in the Imperial State Crown, and stones of 94.45 and 63.65 carats are in the Queen's Crown, along with the notable Koh-i-noor Diamond.

The annals of the South African diamond industry, which today produces 90 percent of the world's diamonds, are fascinating, especially due to the periodic appearance of great stones. The establishment of economic and administrative control over the area, on the other hand, is a highly illuminating chapter in the story of colonialism. In the case of the DeBeers Consolidated Mines—marketing organization for most of the world's industrial and gem diamonds—the elements of the story include vast technical mining problems, great financial battles, amalgamation of small mining companies, the arrival and departure of colorful and strong personalities (Cecil Rhodes, Barnett Barnato, Sir Ernest Oppenheimer), and untold expenditures of funds and labor. For South Africa in general, the diamond continues to be a potent influence in the political and economic life of the country.

Garnet: Garnet is not a single mineral species. It is a whole family of them, including the species grossular, andradite, pyrope, al-mandine, and spessartine, all of which sometimes occur as gemstones. The garnet species are so closely related that among some of them there is complete chemical compatibility. Pyrope and almandine garnets, as an example, may occur in various places with their compositions part pyrope/part almandine in any proportion. All gem garnets are about 7½ in hardness, have specific gravities near 4, and no cleavage, which makes them durable enough for gems. Their crystallographic structures are very much alike, but their color characteristics vary remarkably because of different chemical compositions. Red—in its various shades—seems to be the most common color, so that garnets, like rubies and spinels, have long had mystical significance related to blood. As an example, the Hanzas, while fighting British troops on the Kashmir border in 1892 are said to have used garnet bullets, which they believed would inflict bloodier, more damaging wounds.

If any of the several garnet species deserves the name it would be one of the *almandine-pyrope group.* The word "garnet" comes from "granatum"—the pomegranate—because the gem color was supposed to be the same as that of pomegranate fruit, which has a bright purplish-red flesh. Pyrope is a magnesium aluminum silicate that tends to look deep red to black because it normally contains some almandine to give it color. Almandine is an iron aluminum silicate which tends to be darker, more purplish-red than pyrope. The two together, in their various natural mixtures, give a large series of colors ranging from light red to purplish-red to black. Normally, almandine color is so intense and the stones so opaque that the material is frequently cut in rounded stones—cabochons—with the backs hollowed out to reduce thickness. There is one very popular mix, a variety named rhodolite, that has a very pleasing deep lavender-rose color, occurring particularly in the Cowee Valley of North Carolina. Much larger stones of the same beautiful color have come from Ceylon and, more recently, Tanzania. Another interesting pyrope-almandine, occurring in Idaho, has inclusions of needles appropriately oriented to produce stars.

Almandine-pyrope garnets occur in numerous other places, usually in metamorphic gneisses and schists, but also in igneous granites and in stream gravels. Perhaps the best known of these occurrences is a large area near Trebenice, in the Bohemian region of Czechoslovakia. Much of the gem material used in antique garnet jewelry (which has returned once more to high popularity) came from this source. The Bohemian garnet-cutting industry, which reached its peak in the nineteenth century, tapered off when better material, nicknamed "Cape ruby," was found in the DeBeers, Kimberley, and other South African diamond mines. The Museum of Bohemian Garnets in Trebenice exhibits excellent examples of both antique and modern pyrope jewelry made by local industry. Almandine is common enough to be mined in quantity as an abrasive. Crushed garnet paper is particularly useful for furniture finishing.

The common garnet is the species called *andradite*. It is a calcium iron silicate which, on the whole, is useless for gems except in its black variety, melanite, and the green variety, demantoid. The brilliant luster, good fire, and exquisite grass-green to emerald-green color of this rare gem make it the most highly prized of all the garnets. Naturally, it brings the highest prices. A small amount of chromium substituting in the andradite composition accounts for the color. The finest demantoid gems come from the Ural Mountains of Russia, but they have been found elsewhere. There is a yellow andradite, which unfortunately has been given the confusing name topazolite. It occurs in the Swiss and Italian Alps, but usually, even when good, it is in pieces too small for cutting gems.

The jeweler is not usually familiar with the calcium aluminum garnet by the name *grossular;* he knows it as "hessonite" or as "cinnamon stone," because all gem-quality grossular seems to occur in brownish-yellow, brownish-orange, and brownish-red shades. Pale rose and pale green shades of grossular are very common, but always opaque. Grossular gets its name, unexpectedly, from the Latin word *grossularia* for the gooseberry which it resembles in color in its green varieties. Most gemmy cinnamon garnets come from the gravel beds of Ceylon. Smaller quantities of well-colored stones are found in the state of Minas Gerais, Brazil. The pegmatites of California now and then produce some gem grossular.

Care is required to avoid confusing *spessartine* with grossular. The light orange to brownish-red colors of this manganese aluminum silicate parallel those of hessonite closely. It is used less often in jewelry because production is so limited. Such widely separated lo-

calities as the gem gravels of Ceylon, the pegmatites of Brazil, and the pegmatites of Amelia County, Virginia, have produced superb gems of this species, but not in large quantities. The name comes from its first described occurrence at Spessart, Germany.

Ivory: This is still another of the more widely used and loved gem substances that come from living organisms, so it is not strictly a gem material. For good reasons, ivory and elephant seem inseparably linked in our minds. The ponderous creature, popularized by circuses and zoos, is the principal source for this beautiful, cream-to-white, easily carved, and fairly durable substance. The elephant— like the walrus, hippopotamus, narwahl, sperm whale and boar—supplies ivory through its teeth and tusks.

Ivory is not quite the same substance as bone. Actually, ivory is made of the same kind of dentine which forms our own teeth. About 65 percent of elephant tusk ivory is a hydroxy calcium phosphate. The remainder is a mixture of organic matter consisting mostly of collagen and a little elastin. Because of its somewhat variable composition, the hardness and density of ivory tend to fluctuate considerably. Density of the true ivories varies roughly from 1.95 to 2.0 and hardness from more than 2 to less than 3 on the Mohs scale. Hippopotamus and narwhal ivories tend to be denser and harder than others.

Bone contains a higher percentage of organic material and also tends to be harder and heavier than ivory. After being run through a degreasing process, bone can be used as an ivory substitute. There are decided differences

in appearance between the two materials, however. Their cellular structures are not the same. Ivory is filled throughout with extremely fine channels containing a yellow-brown, jelly-like material. This accounts for its elasticity and the mellow polish that it takes.

Jade: The name jade refers not to one mineral species but primarily to two—jadeite and nephrite—which have practically no relationship to each other except for a similarity of appearance. Jadeite is a sodium aluminum silicate occurring in white, emerald green, and other colors, and is one of a group of rock-forming minerals called pyroxenes. Nephrite is a calcium magnesium iron silicate occurring in colors ranging from white to spinach-green to black, and is one of a different group of rock-forming minerals called amphiboles.

The ancient Chinese worked jade as early as 1000 B.C. The art of jade carving was subsidized by Chinese royalty during the Ching Dynasty (1644-1912), so that it reached its peak in relatively modern times. (Jade and jade carving are discussed in detail in Chapter 10.)

Moonstone: Without having to stretch the imagination too much, the name moonstone fits this beautiful and inexpensive gemstone quite well. Very thin alternate layers of the two feldspar minerals orthoclase (potassium aluminum silicate) and albite (sodium aluminum silicate) cause a light interference effect called "schiller" (see Chapter 2). The bluish glow is very reminiscent of moonlight. In addition to this effect, some moonstone has inclusions appropriately arranged to produce good, if somewhat diffuse, cat's-eyes. Moonstone is usually carved or cut in rounded cabo-

chons with the base parallel to the feldspar layers. It is not faceted because it lacks transparency. Good quality cabochons, however, are translucent and seem to produce their eerie glow from deep inside the stone. Although moonstone has a hardness of only 6 and has a definite cleavage it is a fairly tough gemstone which does not seem to damage easily. The finest rough material comes from Ceylon, either from the gravels of the south or directly from igneous rocks in southern and central regions. There are several other sources, including Madagascar, Burma, India, and even several places in the United States.

Opal: A snide, but revealing, statement was made many years ago by an opal enthusiast. To him it was obvious that "every mineralogist and man of science will rejoice to learn that Queen Victoria exhibits sterling good sense in selecting the opal among her choicest family gifts, thereby presenting a pleasing contrast to the superstitious and foolish fancies of the Empress Eugenie." The statement implied that there was a need to overcome old-time beliefs, still held in high places, about the malevolent influence of opal. At the same time it suggested that opal, as a gem, had plenty to recommend it. Both implications regarding this gemstone are correct.

Opal has had its rise and fall in popularity through the centuries. Its fragile nature —cut gems tend to crack, chip and break easily when subjected to shocks from extreme temperature change or hard knocks—is undoubtedly responsible for some of the adverse feeling about opal. Some opals tend to craze or crack over a period of time even without being stressed. Changing fashions, erratic supply resulting in price fluctuations, and some remnants of the old superstitions all contribute their negative effect on its popularity. And yet much opal is stable, and the incredible beauty of the wide range of brilliant, flashing colors in this gemstone is undeniable.

Most often, precious opal originates in sedimentary deposits. It is not too surprising, then, when fine opal is found as replacements of fossil wood, shells, bone, and even crystals of other minerals. Clustered crystals of glauberite completely replaced by precious opal have been found in Australia, and are popularly known among gem and mineral specimen collectors as "pineapple opal." In Virgin Valley, Nevada, petrified wood is found with a beautiful play of colors from the opal which has replaced it. The story has even been told of the discovery in Australia years ago of a Plesiosaurus skeleton—an ancient sea reptile—completely replaced by opal. Not all the best opal is found as mineral replacements. It also occurs as fillings in seams and cavities in both igneous and sedimentary rocks.

Because the beauty of opal lies in its flashing display of colors, the gem is always cut in cabochon or some other rounded form, which shows the color to best advantage, rather than in a faceted cut. Gem opals are seldom transparent, so that the cutter need not worry about emphasizing brilliance. Facets would only add distracting surface reflections. Opals are not easy to cut, even so. Their tendency to crack if overheated during grinding and polishing operations makes it a gem to be trusted only to a skilled cutter. With a hardness of only

5½ to 6½ the gemstone can be nicely polished, but it must be worn with care or the polished surface is soon dulled by numerous scratches and abrasions.

The first opals available to the gem trade, as far back as Roman times, seem to have come from the mines near Czerwenitza. These mines are now in Czechoslovakia but formerly were in Hungary, so that the opal produced from them has always been called "Hungarian opal." The fortunes of these mines ebbed and flowed with changes from private to government and back to private operation. However, mining activity stopped abruptly with the discovery of the Australian deposits. The milk-white opal with its scatter of pinpoints of color could not compete with the intense, blazing color spectrum of these new opals from "down under."

Australia and Mexico are today the sources of almost all the quality gem opal on the market, with lesser amounts appearing now and then from Brazil. Gems from these sources are different enough in appearance to be distinguished easily from each other. Mexican opal occurs in cavities in volcanic lavas and has a body color which varies from rich cherry red through various shades of orange to colorless. Much of the best quality material is transparent or nearly so, which has a tendency to make the strong play of colors less noticeable. These opals are sometimes backed with a black stone during cutting to give them a better background for the color display. Very likely even the Aztecs, well before the voyages of Columbus, knew the Mexican-type opal. There are a number of occurrences in the states of Chihua-

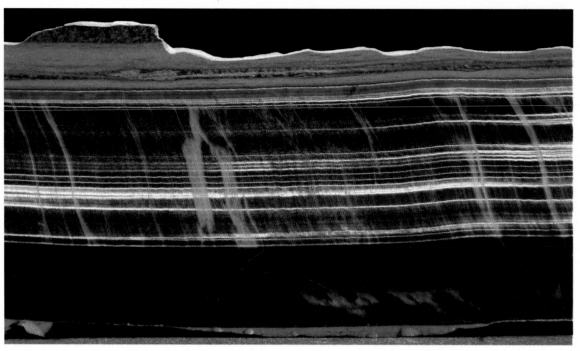

hua, Guerrero, Hidalgo, Jalisco, Michoacan, Queretaro, and San Luis Potosi. The mines in Queretaro, worked intensively for over one hundred years, are by far the most important.

Australian opal, unlike good Mexican opal, is opaque or nearly so, and its body color varies from white through bluish-gray to a dark gray, which is sometimes almost black. The vivid color play, in many hues and often in exquisite patterns, shows strikingly in these stones. The blackish opals are particularly fine and valuable but rather rare, so that often the material is stretched for use by cutting only very thin stones which are then backed with other nonprecious opal or another mineral.

The opal region of south and east Australia is a tremendous spread of uninviting, primitive terrain. Isolation from civilization,

lack of water, and almost unbearable temperatures deter most would-be miners. By the middle 1800's opal had been found in this forbidding "outback" country. In the years since, it has been found at a number of places in the states of New South Wales, South Australia, and Queensland. More discoveries are certain to be made now that all the early resistance to acceptance of this wondrous material in the gem market has vanished. The names Yowah, Quilpie, White Cliffs, Lightning Ridge, Coober Pedy, Eulo, and Tintabar—all famous Australian opal occurrences—add romance to the discussion of opal, just as Kimberley and DeBeers do to diamond.

Pearl: Edwin Streeter, English gem dealer, wrote a century ago about the Roman greed for pearls: "Pompey, the victorious

Roman general, found in the palace of Mithridates a wonderful collection of precious pearls. . . . In his third great triumph against the Asiatic princes in 61 B.C. he took thirty-three crowns of pearls. After this period the pearl luxury became quite a disease in Rome. The philosopher Seneca spoke very sharply against the Roman women for wearing so many pearls. . . . Roman ladies wore necklaces of pearls, also ornaments for the breast consisting of thirty-four half-ball pearls . . . dresses, shoes and bracelets richly covered with costly pearls." Julius Caesar, who collected art objects and jewels during his military campaigns, was especially fond of pearls. It is even suspected that his invasion of Britain was started with the hope that he would find a supply of freshwater pearls there. For perhaps two thousand years rather beautiful pearls have been found in the rivers of Scotland, Ireland, and Wales.

The greater part of the story of gem pearls centers on the sea as salt-water environment for a number of different kinds of host mollusks. Freshwater pearls, although they may occasionally be lovely to look at and well formed, do not have the fine orient, or luster, of salt-water pearls. Most of the inland pearls of Europe and America are formed in one of the mussels of the genus Unio, but any shell-secreting mollusk is capable of producing pearls of one sort or another. Without doubt, most of the finest natural pearls in the gem trade, those with an incomparable luster and bearing the highest prices, come from the salt waters of the Persian Gulf. Some equally fine pearls come from the Gulf of Manaar, which lies between Ceylon and India. At both of these groups of fisheries the pearl-bearing mollusk is *Pinctada vulgaris*. The only other major pearl-fishing area is in the Red Sea, and there the same mollusk species has provided fine pearls back to the time of ancient Egypt. Natural salt-water pearls are still valuable, but production has dropped off to nothing in most of these established fisheries with only minor amounts coming from others. Overfishing and destruction of the mollusk beds have taken their toll.

The protective shell secreted by a pearl-producing mollusk is rather rough on the outside. On the inside, nature has made protective provisions by arranging for a process of secretion by the animal's mantle that results in a smooth surface against which the mantle can bear without excessive irritation or damage. This inner layer is known as the nacreous layer and the nacre of which it is composed is built of tiny, flat, overlapping plates of calcium carbonate. It is the soft sheen of light reflected from this platy layer that is so highly prized. Known popularly as mother-of-pearl, it is the same material which goes into the formation of pearls. The shell's nacre lining was designed for protection against irritation, and pearl formation is merely an extension of this process.

Should some irritating piece of debris, such as a small piece of shell or a sand grain, accidentally breach the mollusk's sanctuary and get inside the shell the animal will force it against the inside of the shell and slowly plaster it over with layer upon layer of nacre. This accounts for the formation of blister pearls. Sometimes the invader is a wriggling parasite, or a too-elusive fragment, so that it resists being plastered over. In self-protection, the mollusk develops a cavity which slowly engulfs the irritant in a cyst or pocket within its body. Totally surrounded by mantle tissue, the secretion process proceeds against the irritant until it is buried many layers deep in nacre to form a more or less spherical pearl.

Since pearl is composed of calcium carbonate, in the form of the mineral aragonite, it should have most of the characteristics of this mineral. However, pearl is really a mixture of aragonite with an organic cementing substance called conchiolin as well as a small amount of trapped water. Its density, then, varies from 2.6 to 2.78, and its hardness from $2\frac{1}{2}$ to $3\frac{1}{2}$. Obviously, because of its composition and properties, pearl is easily damaged. In spite of this, a pearl necklace can be made to last almost forever with appropriate care. The major problems are cracking and dulling due to excessive drying of the organic conchiolin and the attack of weak acids. Prolonged exposure to hot and excessively dry air should be avoided, and periodic cleaning to remove skin acids will prolong beauty indefinitely.

Shape is one of the important factors in determining the value of a pearl. Others are size, luster, surface quality, and color. Spheres and drop-shaped pearls are most highly prized, but even the more common, oddly formed "baroque" pearls are useful in jewelry. Discolored spots, as well as small ridges and pits on the surface, are most undesirable. Sometimes the value of such a marred pearl can be increased or restored by carefully peeling away the imperfect surface layer, but the process is not certain or practical.

Luster is very important. It arises from a combination of diffraction and reflection of light by the tiny plates of the nacreous layers. The more translucent the body of the pearl may be, the better its attractive iridescence.

While luster is the prime consideration for beauty, the size of the pearl goes far in determining its price. The gradations are as follows: very small, 3 millimeters or less; small, 3 to $4\frac{1}{2}$ millimeters; medium, 5 to 6 millimeters; large, 7 to 8 millimeters; very large, 8 millimeters or more.

Pearl colors are delicate and highly variable. Usually pearls are envisioned as being white or near white. Actually, they come in many colors and subtle shades, with blackish, golden-yellow, pink, creamy-white, and white being considered most desirable.

However desirable, rare, expensive, and exotic the natural pearl may be, it is not particularly significant in the modern gem-pearl market. The Japanese cultured pearl, produced by careful implantation and large-scale farming of the species *Pinctada martensii*, reaches the market by the millions each year. It has all the beauty and desirability of a totally natural pearl and is practically indistinguishable from it. A detailed account of the cultured pearl and its production is given in Chapter 6.

Peridot: A story persists that in certain parts of Arizona and New Mexico the ants have mined and brought to the surface large numbers of pebbles of peridot to build their ant hills. Fortunately, the story is not true and the peridot pebbles are there because they are a basic part of the local soils. The lovely peridot deserves better than an ant heap. To some people, the green color of this iron magnesium silicate, which ranges from yellow-green to deep bottle-green, is the most attractive of all the gem colors. There is a considerable variation in color because of variable amounts of iron and magnesium present. Peridot is an ancient gem name for the mineral series now known by the name olivine. The olivine series varies in composition from an iron silicate called fayalite to a magnesium silicate called forsterite. Usually, samples of olivine of any composition range from opaque to partially translucent. Rarely are they transparent, as in the gravels of Burma and the metamorphosed, nickel-bearing rocks of the island of Zebirget in the Red Sea. All the fine peridot gems of modern times come primarily from these two occurrences. The beautiful, deep green, 310-carat, record-size gem in the Smithsonian collection is from Zebirget—formerly known as St. John's Island and, in ancient times, Topazios. All the occurrences of peridot can be traced back to two kinds of igneous rocks called basalts and peridotites. Strangely enough, it is an important constituent of many meteorites.

Many of the ancient "emeralds" and many of the presumed emeralds gracing royal treasuries are really peridot. It is difficult to understand the confusion, because their colors are decidedly different. Even in recent times the confusion has been compounded, for peridot was long known by poetic license as "the evening emerald." Despite the attempts to glamorize this gem by association with emerald it can pretty well stand on its own merits. Certainly its color is attractive. A hardness of 6½ is adequate, and a tendency to cleave under excessive stress can be overcome by mounting it in supporting metal jewelry. The gem shows very little fire and has a tendency toward an oily and somewhat glassy luster. Its refractive index, fortunately, is high enough to make the gems attractively brilliant.

Quartz: This mineral is cursed, unfortunately, with so many names for its multitudinous varieties that it is perhaps the most confused and confusing of the gemstones. In a way, this is not too surprising. Silicon and oxygen comprise most of the earth's crust. Naturally, then, the compound silicon dioxide, or quartz, is found in almost all the rocks of the crust. Being so widespread and forming in so many different chemical and physical environments, quartz should be expected to occur in numerous colors, forms, patterns, and impure mixtures. Many of the varieties, although common, are remarkably attractive and have been assigned special gem names at various times in the past. Fortunately, most of these names never achieve widespread acceptance and fade from common usage. A few, because of the sheer beauty of the material with which they are associated, have persisted. These include amethyst, rock crystal, citrine, rose quartz, smoky quartz, agate, jasper, onyx, bloodstone, carnelian, chrysoprase, and others.

For convenience, and by custom, the gem-quartz varieties are divided into two groups: crystalline, and fine-grained, or cryptocrystalline, quartz. Typical crystalline varieties—including amethyst, rock crystal, citrine, rose quartz, and smoky quartz—occur in clear, clean, transparent pieces sufficiently large to cut into faceted stones. Typical cryptocrystal-

111

line varieties—including agate, jasper, onyx, bloodstone, carnelian, and chrysoprase—are actually such fine-grained masses of tiny crystals that they tend to be translucent or opaque, although beautifully colored and patterned.

Whatever the color, pattern, or transparency, all quartz has a hardness of about 7, with some of the cryptocrystalline varieties tending to be slightly softer. This degree of hardness is adequate for gem purposes, but the mineral has very little fire and a vitreous or glassy luster. Except when well colored or beautifully patterned, it has little to offer as a gemstone.

The name *amethyst* invokes images of a purple gem, yet this stone has a remarkable color range. From Madagascar and Hungary come bluish-violet stones, from North Carolina and Guanajuato, Mexico, come reddish-violet tints. Some Brazilian stones have brownish-red overtones in the purple, while those from southeast Pennsylvania may be a smoky purple. Unfortunately, in all occurrences the color tends to be spotty in the crystals. Sometimes only portions of a crystal—usually the tip—tend to be amethyst-colored with the remainder colorless or smoky. A skilled gem cutter can orient a finished cut gem so that even a spot of amethyst color can transmit its effect to the entire stone. Regardless of the sources of amethyst, the gem trade usually classifies the most desirable, purple-colored stones as "Siberian," more poorly colored stones as "Uruguayan," and the least desirable as "Bahian." Reasons for the coloration of amethyst are not definitely known, although the presence of a very small amount of iron contamination seems to have much to do with

it. Whatever the reasons, it is a popular color, and amethyst is the most coveted gem variety of quartz. Should there be a return of the ancient belief that amethyst guards against intoxication—its name is from a Greek word meaning "not drunken"—the gem will probably rise even higher in popularity. More than likely the name originally referred only to the fact that amethyst color approaches, but is not the same as, that of a good red wine.

Citrine will be described in any text on gems as a yellow to golden-brown to red-brown variety of quartz. What most texts fail to mention is that quartz in these colors is quite rare in nature. Almost all citrine in the gem market today is prepared by the careful heat treatment of certain varieties of amethyst at between 500 and 600 degrees centigrade. (Strangely enough, certain kinds of Brazilian amethyst turn grayish-green to grass-green when heated.) Further misunderstanding arises because heat-treated citrine is often sold under such names as Madeira topaz, Spanish topaz, or even topaz. At the same time, true topaz is sometimes sold as Brazilian topaz. Thus, the two distinctly different gems tend to become indistinguishable in the minds of potential gem customers. True topaz of the yellow citrine color is, of course, very rare and much more valuable.

Rock crystal is the stone of the crystal gazer, and any such teller of future events who owns a good, clear, flawless sphere of quartz as much as 6 inches in diameter is fortunate, indeed. Although rock crystal, or colorless quartz, is very abundant and may even occur in single crystals weighing several tons each, large and flawless pieces for cutting are very rare and thus terribly expensive. There is an extraordinary, flawless sphere in the Smithsonian collection which is 12⅞ inches in diameter and weighs 106¾ pounds.

Obviously, if such fine spheres can be cut from rock crystal, it must be very desirable for general rock-carving purposes. Rock crystal is cut and carved into faceted stones, beads in various patterns, and many other objects from handles to chandeliers. The material is so common and inexpensive that almost the total value of such objects is in the cost of cutting and merchandising. The art of carving and engraving rock crystal has been well developed because the material lends itself so well to this treatment. For centuries, Asians have carved superb objects, and modern gem carvers everywhere find rock crystal to be an excellent material for their art and craft work. Brazil has been the best source of gem rock crystal for more than a century. Individual crystals weighing up to twenty-five tons have been found there. The largest recorded crystal—about twenty feet long and several feet thick, weighing more than forty-four tons—was found in the state of Goiaz. Madagascar has also been an excellent source of rock crystal, having supplied carvers in India, Europe, and China as far back as the ninth and tenth centuries.

Rose quartz occurs in delicate shades of rose-pink to rose-red, sometimes with soft lavender overtones, and is among the more desirable quartz varieties. Rarely does it occur in crystals or in large enough pieces with the clarity needed for faceted gems. This is unimportant because it shows its color best anyway when carved or cut *en cabochon*. When avail-

able it is one of the more popular carving materials. Happily, rose quartz sometimes carries quantities of fine needle inclusions of rutile. They do not have much effect on its color, but may produce fairly strong asterism when properly oriented. The stars, oddly enough, are best seen by looking at a light through the bead or sphere.

Rose quartz usually occurs in the central quartz core of pegmatites. Such a pegmatite core of rose quartz is exposed near Custer, South Dakota, where it is visible as an enormous pink mass more than one hundred feet long. This mass was the source of material for Chinese carvers for many years until it was replaced by better-quality Brazilian material. Brazil, then, is the primary source for this attractive variety of quartz, with the bulk of it coming from the Arasuahy-Jequitinhonha District in Minas Gerais. There are, of course, several other occurrences in places from Maine to California and on around the globe. In recent years some remarkable crystal groups for mineral specimen collectors have been found in Brazil.

Smoky quartz has a wide color range. Specimens occur in very light brown and in shades of brown increasingly intense, so that the quartz looks black. Any of these color grades may also have smoky overtones, and may at times look more gray than brown. The mineral typically forms in high-temperature deposits such as pegmatites. Cairngorm is a kind of smoky quartz from the Cairngorm long been used to decorate the traditional Scotch Highland costume. Most of the Scotch Mountains in the Scottish Highlands. It has

113

Above: Oddly cut but excellent 177-carat gem kunzite from California. Left: 9-carat spinel from Burma in preferred ruby color. Right: 77-carat Berman Topaz from Brazil. Opposite: One of relatively few preserved, well-formed gem crystals of California kunzite (one-half natural size).

supply now comes from Brazil. As with other quartz varieties, smoky quartz occurrences are scattered around the world. The best of them all is perhaps the Swiss Alps, which have at times yielded large tonnages of superior material. There is some doubt about the cause of color in smoky quartz, as there is with several other varieties of quartz. Almost certainly there is a relationship between the intensity of its coloration and the amount of natural radioactivity to which it has been subjected.

Chalcedony and *agate* are terms often used interchangeably for fine-grained varieties of quartz. Usually the name agate is reserved for those kinds of chalcedony which show some recognizable pattern of banding, marking, or coloration. There are many of them which do.

For gem purposes, perhaps the most valuable of the chalcedonies is *chrysoprase.* The best of this material, found in Australia, is translucent and quite uniform in its bright grassy-green color. Cut into cabochons or beads which take advantage of these color- and light-transmission characteristics, good chrysoprase is as beautiful—and almost as expensive—as good quality jade. Certainly it is far more attractive and durable than serpentine, which is often used as a jade substitute. *Plasma* and *prase* are also green varieties of quartz. They lack the intensity of color and the beauty of chrysoprase because it alone acquires its color from the presence of nickel compounds. The others get a less rich green color from the presence of iron compounds.

When the coloring impurity is hematite or red iron oxide, the chalcedony is called *carnelian.* Other iron oxides produce yellow-ish to brownish-red colors. These are called *sard,* and if the color alternates in bands it is called *sardonyx.* Black and white banded chalcedony is called *onyx.* The unfortunate aspect of all the colored chalcedonies is that their bright hues can be induced by treatment with appropriate dyes. True carnelian, for example, is found in several places, and in some quantity on the Brazil-Uruguay border. Most of it in commerce, however, is carefully dyed chalcedony which originally was too pale or colorless. This is also true of black onyx. The process in this case is rather interesting. First the stone is boiled in a sugar solution and then soaked in concentrated sulphuric acid. Black carbon, formed by the acid destruction of the sugar solution which has soaked in, is deposited in the porous quartz.

The *patterned agates* are perhaps the most interesting of all. There seems to be no end to the permutations and combinations of color and design. Moss agate and plume agate derive their delicate, plant-like patterns from branching or dendritic impregnations of other minerals in the chalcedony. These indeed, are, often mistaken for plant fossils but have no relationship at all to living materials. Montana, Wyoming, Oregon, and Texas in the United States, and localities in India, Uruguay, and Brazil have produced superb agates of this type. Banded or fortification agates may have their parallel-line patterns all running as straight lines in one direction, or may actually show concentric banding so uniform that they look like target bull's-eyes. Big tonnages of excellent banded agates in many colors and patterns are coming into the gem-cutting market these days from Mexico and Brazil. Fancy agate collecting has now reached its peak as a hobby in the western United States with thousands of devotees involved in uncovering new types and new sources. New varietal names are constantly being introduced to distinguish newly discovered patterns: eye agate, fire agate, iris agate, lace agate, crazy lace, lavender lace, Lake Superior agate, Apache plume, etc.

There are other groups of quartz family gems that should at least be mentioned. Certainly *jasper*, a quartz containing a high percentage of impurities, is one of them. Its color varies from a rather bright red-brown to the well-known bloodstone type which is green with red spots. There is also petrified wood of various genera and species, from palms to oaks to conifers. It often assumes the attractively detailed pattern of the original wood now completely replaced by quartz.

As mentioned earlier, there are innumerable names applied to the varieties of quartz. The long but partial list given here illuminates the wide range and long history of quartz among the gemstones. Quartz has been called agate, agate jasper, Alaska diamond, Aleppo stone, amberine, amethyst, amethystine quartz, Ancona ruby, apricotine, Arkansas diamond, aventurine, azure quartz, Baffa diamond, basanite, beckite, beekite, bishop's stone, bloodstone, blood jasper, blue chrysoprase, Bohemian diamond, Bohemian topaz, Bohemian ruby, Brazilian diamond, Brazilian topaz, Bristol diamond, burnt amethyst, Cairngorm, California moonstone, Cape May diamond, carnelian, carnelian onyx, Catalinite, Catalina sardonyx, chalchihuitl, chalcedony, chalcedony onyx, chalcedonyx, chert, chinarump, chrysoprase, Cornish diamond, cotterite, creolite, crispite, crystal, cupid's darts, Dauphine diamond, dendritic agate, eldoradoite, emeraldine, enhydros, eye agate, false diamond, false lapis, false topaz, *flèches d'amour,* flint, flower stone, fortification agate, frost stone, gold quartz, golden topaz, hairstone, heliotrope, Herkimer diamond, Horatio diamond, hornstone, Hot Springs diamond, hyacinth, Indian agate, Indian topaz, iolanthite, iridescent quartz, iris, Irish diamond, jaspagate, jasper, jasperine, kinradite, lavendine, love arrows, Lydian stone, Madeira topaz, milky quartz, mocha stone, Mohave moonstone, Montana agate, Mont Blanc ruby, Mora diamond, morion, moss jasper, mother-of-emerald, myrickite, needle stone, nicolo, novaculite, occidental agate, occidental amethyst, occidental cat's-eye, occidental chalcedony, occidental diamond, occidental topaz, onegite, onyx, orange topaz, oriental agate, oriental chalcedony, oriental jasper, ouachita stone, Pecos diamond, petrified wood, plasma, prase, prismatic moonstone, pseudo-diamond, Quebec diamond, rainbow agate, rainbow quartz, rhinestone, riband agate, riband jasper, ribbon agate, ring agate, river agate, rock crystal, rose quartz, rubasse, sagenite, sandy sard, sapphire quartz, sapphirine, sard, sardoine, sardonyx, Schiller quartz, Scotch topaz, semicarnelian, Siberian amethyst, siderite, sinople, silicified wood, smoky quartz, smoky topaz, soldier's stone, Spanish topaz, St. Stephen's stone, star stone, starolite, Swiss lapis, test stone, Texas agate, Thetis hairstone, tiger-eye, touchstone, tree agate, tree stone, Trenton diamond,

117

unripe diamond, Venus hairstone, violite, water agate, wax agate, white carnelian, wood agate, zonite, and many other names.

Spinel: Ruby, to the eye, looks very much like some spinel, but most spinel doesn't resemble ruby at all. Only the red and deep pink varieties of spinel have this honor. Mention has already been made of the historic Black Prince's Ruby and the Timur Ruby, both of which are actually spinel. Very likely much of the ruby in the regal jewelry of the medieval western world was and is spinel. Survival of the name "balas ruby" for red spinel, even into modern times, indicates something of the confusion in identification that once existed. The mineralogist Rome de Lisle, in 1783, is credited as the first scientist to distinguish clearly the differences between true ruby and red spinel.

Part of the reason for the lesser popularity of spinel is undoubtedly because its colors —red, mauve, blue, burnt orange, red-orange, brown, lilac, grayish-purple, violet, greenish, wine-red, steel-gray, gray-blue, slate, indigo, rose-brown, and black—make it resemble too many other gems.

The mineral is not particularly rare, but rough spinel large enough to cut into good stones weighing over 10 carats is unusual. However, top quality gems in sizes up to more than 100 carats have been cut from material found in the gem gravels of Burma and Ceylon. There are reports of large spinels being imported into England from India in the middle 1800's, including one weighing 197 carats. The two gravel-bed occurrences mentioned produce most gem spinel today, even though it is found in many other places such as Thailand, Australia, and Brazil. Wherever it is found in place, instead of in stream gravels, it generally occurs with corundum as a constituent of gneiss, limestone, and other metamorphic rocks.

Spinel is magnesium aluminate with a somewhat variable composition, since iron, maganese, chromium, and zinc can enter into it by substitution for the magnesium and aluminum. Normal gem spinels, whatever their composition, have a refractive index just under 1.72. They are brilliant enough, but have a rather low light-dispersing ability, which gives them little fire. The gem has a beautiful suite of colors and is quite hard—at 8 it is as hard as topaz. Spinel occurs in remarkably flawless crystals and pebbles: it is surprising that it is not valued more highly.

Spodumene: There is no ancient history for gem spodumene. The mineral has been known to mineralogists as a lithium aluminum silicate for at least one hundred years. It had no importance as a gemstone during the couple of centuries it was known as a mineral until small amounts of the bright green variety, hiddenite—discovered by W. E. Hidden—were found in North Carolina about 1879. Unfortunately, not enough of it was recovered to make any impression on the gem trade. Spodumene really dates its acceptance from the time of the discovery of a good quantity of a beautiful amethyst-pink to lavender-rose variety at Pala, San Diego County, California, in 1903. This new variety was named to honor George F. Kunz, gemologist, who first described its characteristics. Both kunzite

119

and hiddenite exhibit strong pleochroism. Depending on the direction of view through a crystal, fragment, or cut stone, kunzite from California is pinkish-lavender, violet, and colorless, while that from Brazil is violet, purple, and greenish. Hiddenite is bluish-green, yellowish-green, and emerald-green. It is most important for the lapidary to cut gems with the proper orientation in the material to capture the maximum intensity of color. All spodumene has a hardness of 7, a refractive index of about 1.66, and shows very little fire. Cut stones are difficult to form, except for the experienced lapidary, because the mineral has a tendency to cleave easily.

Lithium, a major part of the composition of spodumene, is common enough in pegmatites, so that the mineral is not rare. As a matter of fact, nongem crystals as much as forty feet long have been found at the Etta Mine in the Black Hills of South Dakota. Even gem spodumene is not so rare as once it was presumed to be. Large quantities of kunzite were found all at once in Brazilian pegmatites in the 1960's. Regularly, smaller quantities of greenish and yellowish gem crystals and fragments of spodumene are recovered from pegmatites in Brazil, Madagascar, and elsewhere.

Topaz: Over a period of time the facts about topaz and its identity have suffered from much confusion. Even its name has been badly chosen, because there is a strong possibility that to the ancients the word was applied to an entirely different gem, what we now call peridot. However, the name "topaz" began to come into regular use in the late 1700's for an aluminum fluosilicate with a hardness of 8, and a disagreeable tendency to cleave markedly and easily. It was recovered in some quantity from the mines at Schneckenstein, Saxony. For about seventy years, ending in 1797, this topaz mining venture was a royal enterprise. During this time, about 150,000 grams of the pale, wine-yellow gemstone was recovered. Unfortunately, from the start, topaz came to be known as a yellowish to orange-yellow gem. This caused double confusion. First, there are several other colors of topaz. Second, other gemstones of similar color, such as yellowish quartz, flood the market and are sold under various names suggesting that they are somehow related to true topaz. Meanwhile, blue, pink, brown, or colorless topaz is suspect because it carries the proper name but is not yellow.

Most topaz is colorless. From Brazil have come beautifully formed, enormous crystals of this kind weighing several hundred pounds each. Naturally, existing in such quantity, colorless topaz has minimum market value as gem material. Sherry or muscatel wine-colored stones, or those with natural bright pink or reddish-purple tones, are very highly prized. It is possible, by heat treatment of certain pale yellow or brown stones, to convert them to good pink or blue hues. Some of these heat-treated blue stones and natural blue stones, too, resemble good aquamarine and can easily be confused with it unless they are carefully examined.

Most topaz comes originally from pegmatites, although much of it is found as water-worn pebbles in gravels that have been eroded from these rocks. The mineral species is not particularly rare; deposits of it are found scattered in almost every corner of the earth. The most important commercial deposits are scattered over a large area near Ouro Preto, Minas Gerais, Brazil. Blue, white, yellow, and colorless varieties of topaz are found elsewhere in Brazil. Texas, California, Colorado, New Hampshire, and Utah have all provided at least a small amount of gem-quality topaz through the years. Notable amounts are found in the gem gravels of Ceylon and Burma, which are so productive of many other gem species. The list of interesting occurrences should be expanded to include mention of the fine gems from the Ural and Ilmen Mountains in Russia, Klein Spitzkopje in Southwest Africa, northern Nigeria, and so on. This ready availability, coupled with the difficulties of cutting and wearing a gem that cleaves so well under shock, has helped keep the price of topaz low despite its superior hardness and often beautiful color.

Tourmaline: Considering that the chemical composition of tourmaline is so variable, it is no surprise to learn that the gemstone is found with a great color range. Its hues include pink, green, blue, yellow, brown, black —all in many different shades and combinations of shades. Tourmaline with a color near emerald-green is particularly popular. Chemically, this gemstone is a very complex borosilicate and its color is determined by variable amounts of the elements in it. Tourmaline crystals having sodium, lithium, or potassium are either colorless, red, or green; those having iron are blue, blue-green, or black; and those having magnesium are colorless, yellow-brown, or blackish-brown.

The actual origin of the gem name is lost, but it is believed to come from the Sinhalese word *turmali*. This name was originally applied in Ceylon to a yellow variety of zircon. The tale is told that a package of yellow tourmaline, labeled *"turmali,"* was sent to the Dutch gem markets in 1703. Though the mineral was mislabeled, the name has stayed with it into our own times.

Tourmaline is found in both igneous and metamorphic rocks. In general, it can be said that those varieties containing magnesium occur in metamorphosed rocks, such as schists and limestones, where they occur with the minerals quartz, biotite, phlogopite, augite, plagioclase feldspars, tremolite, and others. The other varieties are usually found in pegmatites associated with quartz, albite, orthoclase, microcline, muscovite, lepidolite, beryl, spodumene, and others.

Tourmaline has other satisfactory characteristics, in addition to good color, which make it an excellent gemstone. Hardness varies a bit, depending on composition, but it averages at a suitable 7¼. The mineral has no tendency to cleave and, although its luster tends to be somewhat glassy, it oftens has a high degree of transparency which shows the color well.

There are two exotic characteristics of tourmaline that are worth special mention. One is its usually strong dichroism. Crystals or fragments of the mineral show much stronger color when viewed in the direction of the long crystal axis than when viewed from any other direction. This means that if the crystal is

122

Opposite: Three fine tourmalines—118 carats from Brazil (top), 28 carats, also from Brazil (bottom left), 21 carats from Maine (right). Below: 98-carat Ceylon zircon. Bottom left: 25-carat rhodolite garnet from Tanzania. Right: A topaz of 34 carats from Brazil.

dark, the cutter will have to cut the stone with the flat part, or table, parallel to the long axis of the crystal. Similarly, the table of a lighter-colored crystal can be cut perpendicular to the long axis in order to produce a deeper-colored gem. The other noteworthy characteristic is electrical. It is the direct result of the non-symmetrical structure of tourmaline when examined from end to end. One end of a tourmaline crystal is always different from the other. When crystals or crystal fragments are heated or compressed, an electrical charge difference develops on the two ends, which accounts for the dust-attracting property of tourmaline gems that are enclosed in lighted and closed showcases.

Our best tourmaline gemstones these days come from Mozambique, Madagascar, and Brazil. In the past there have been excellent discoveries at Mursinka, Nerchinsk, and elsewhere in Russia, in the gem gravels of both Burma and Ceylon (Burmese red stones are considered very desirable), and in Africa at various places, including Tanzania, South-West Africa, and Southern Rhodesia. In the United States some of the finest tourmaline gems known came from Mt. Mica, Maine, in 1820. A few Maine stones are still found now and then. Later a flood of fine tourmalines of almost every color came from San Diego County, California. These are still being mined, but generally do not compete with stones from the chief producing localities in Africa and Brazil.

Turquoise: Without its color, turquoise would have almost nothing to offer as gem material. It is a dull, opaque mineral, but with a color so distinctive that "turquoise" has become one of the standard shades of blue. The name is derived from the old French *pierre Turquoise,* or stone of Turkey. None of the material is mined in Turkey, but most old-time turquoise was sold in Turkish markets.

Most of the turquoise mined today is just about worthless because of its inferior color. Some of this quantity is salvaged by impregnation of the porous material with various plastics to improve its physical quality. The addition of a blue dye can easily mask a natural greenish cast or add intensity to a weak blue. Essentially, the mineral is a hydrous copper aluminum phosphate, which gets its color from the copper in its composition. Iron often influences the shade of blue, too. Some turquoise tends eventually to fade or become greenish, but a stone which has been stable for some time will usually stay that way unless maltreated by excessive heat, exposure to oils and dirt, and unnecessary abrasion. A hardness slightly less than 6 makes turquoise one of the softer gemstones, so that it needs greater care and protection than most. Although opaque, the stone takes an excellent polish, so it is used almost exclusively for making cabochons, beads, and carvings.

To supply the demand there is continuous mining activity. Unquestionably, the finest material still comes from the Nishapur district in Iran, as it has since ancient times. There it occurs with limonite as a filling in rock which has been badly broken up and weathered. In the United States large quantities of turquoise have been mined for hundreds of years from a similar kind of rock. The material from the

Los Cerrillos Mountains and elsewhere in the Southwest is generally inferior to Persian turquoise. However, it too has remained popular and is the basis for a thriving Indian jewelry industry. Usually it is mounted in silver in traditional and modern designs.

Zircon: Golden-brown, sky-blue, and colorless gems of zircon are circulated widely in the gem trade. Considering the high quality and beauty of these gems it is surprising that they have never become generally popular. A high refractive index and a high dispersion give zircons considerable brilliance and fire, so much so that colorless stones can even be used as satisfactory substitutes for diamond. Of course, close examination gives away the deception because zircon, being a doubly refracting substance, shows a doubling of the edges of the back facets when viewed through the cut gem. Diamond, being singly refracting, does not produce this edge-doubling illusion. There is a gross difference in hardness, too, with zircon varying between 7 and 7½ compared to a much harder 10 for diamond.

The colorless, blue, and golden-brown gems that are the most popular and best known were all found as reddish-brown pebbles in the widespread gem-gravel fields of Indochina. Careful heat treatment at temperatures near 1800 degrees Fahrenheit for periods of time up to two hours causes a color change to gold or blue, depending on how the air supply is controlled in the crude heating ovens. The new color is usually quite stable, but in some cases has been known to revert in time to unattractive shades of green or gray-brown.

Zircon is a zirconium silicate containing a little iron impurity which replaces part of the zirconium. The elements hafnium, thorium, and uranium are also present, in some cases making up as much as 4 percent of the total. Thorium and uranium are radioactive elements and, if present in any quantity, subject the structure of the zircon to a constant bombardment of radioactive particles. In time, this bombardment may actually disarrange or destroy the structure, so that the material is still zircon by composition, but has no zircon structure. Minerals having such broken-down structures are called *metamict* minerals. It is obvious that all sorts of stones will exist in which there is a variable amount of structure derangement, depending on how long the process has been going on. This means that the characteristics of a particular gem zircon—such as hardness, specific gravity, etc.—may be anywhere between those of completely crystalline and completely metamict material. Metamict zircon can usually be identified by its medium to dark-green color.

Zircon is quite common and is often found as an accessory mineral in igneous rocks. Gem zircon is not so common. As is true of many other gems, this one is found in quantity in the gem gravels of Ceylon and Burma. By far the most important producing areas lie in Indochina in a region comprising parts of Viet Nam, Laos, and Thailand. Bangkok is the most important cutting and marketing area for these stones. War conditions in Southeast Asia have made supplies erratic, especially of the best material for heat-treating, which comes from the easternmost part of the region.

Stones for Collectors & Craftsmen|5

Commercial gems are considered an elite group, singled out as being more "precious." The other stones comprising the gem kingdom are relegated to other, lesser categories such as "unusual," "semiprecious," "nonprecious," or "decorative" stones. This seems to imply that they are somehow less good. This may actually be the case; perhaps some catastrophic weakness such as softness, or easy cleavage, or even excessive rarity, eliminates the gem as serious competition for the select few. However, some of these stones may be just as valuable as their honored cousins, or may be useful for purposes to which the "precious" stones are unsuited.

Fortunately for gem lovers with collecting instincts there is a remarkable array of mineral species which can supply cut gems of some interest other than commercial value. Many of them are of transparent, faceting quality. Many more have some special desirable color or characteristic or scarcity which stimulates competition among collectors. With increasing demand for colored gems in the jewelry trade, there is a tendency for some of the odd gems to be promoted into commercial channels. Sometimes a new supply of some exotic gem material creates an active market for it. Now and then the reverse process applies and a strong market interest in some particular gemstone stimulates prospecting and the development of new supplies.

There is yet another class of mineral materials, such as rhodochrosite, malachite, and serpentine, that hold a place somewhere among the gemstones by virtue of their beauty when worked into objects of massive size. Unsuited for expensive jewelry and generally lacking in most of the important gem attributes, they often find ideal use as carvings or other art objects or as decorative elements in the architecture of buildings.

This summary of the "lesser" gemstones considers first those with appeal for collectors and, second, those of primarily ornamental value.

Gem Collectibles

Amblygonite: This is a lithium aluminum fluophosphate sometimes found in transparent, gem-quality crystals. It varies from colorless to yellow, has a hardness of 6, and tends unfortunately to develop its perfect cleavage if stressed too much. Gem-quality amblygonite is found mostly in Sao Paulo and Minas Gerais, Brazil.

Andalusite: Named for Andalusia, Spain, where it was first found, this species is an aluminum silicate with a hardness of 7½. Its color varies from green to greenish-brown. Most of the gem material comes from the states of Espirito Santo and Minas Gerais in Brazil, where it is found in rounded pebbles.

Apatite: Although it is a very common mineral occurring in enormous crystals and at innumerable places around the earth, apatite is seldom found in transparent, gemmy crystals or fragments. The mineral is a calcium phosphate containing fluorine or chlorine and frequently other elements. A hardness of only 5 and its brittleness make it useless for ordinary gem uses, even though it occurs in a wide variety of good colors. Cerro de Mercado in

Mexico produces finger-size, golden-yellow, transparent crystals. Mogok in Upper Burma has yielded good blue stones. Yellowish-green, violet, and deep-blue stones have appeared elsewhere and even apatite cat's-eyes are found in Ceylon.

Augelite: A colorless, hydrated aluminum phosphate with a hardness of only 5, augelite is a collector's item only because of its rarity; it has no particularly attractive or unusual features. Cuttable crystals have been found in the United States and in Bolivia.

Axinite: This is usually found in dark cinnamon to greenish-brown crystals. Their shapes are reminiscent of an axe head, which accounts for its name. It is a chemically complex calcium aluminum borosilicate and is usually too dark in color to cut into attractive gems. When suitable light-colored material is found for cutting, the hardness of 7 gives it sufficient durability to make it worthwhile. Cutting-quality crystals have been found in Switzerland, France, and Baja California.

Benitoite: In the gem collection of the Smithsonian Institution is a 7.5-carat, beautiful blue, cut stone that resembles a sapphire, but with exceptionally good fire. The fire is actually better than that of diamond, but is subdued by the blue color. The stone is benitoite, a very rare barium titanium silicate named to honor San Benito County, California, the only place it has ever been found. The 7.5-carat gem is the largest cut stone in existence for this species. A hardness of 6½ gives it satisfactory durability. With all its excellent qualities, it cannot be a commercial gem because it is so rare.

Beryllonite: The best beryllonite of gem-cutting quality, a very rare beryllium sodium phosphate, has been found at Stoneham, Maine, with other pegmatite minerals. Like augelite, it is colorless, looks glassy, has a hardness of only 5, and offers little except its great rarity.

Brazilianite: A pegmatite in Minas Gerais yielded in 1944 some superb, large gem crystals of a hydrous aluminum sodium phosphate previously unknown to science. It is a brittle substance with a hardness of only 5½ and is otherwise almost undistinguished as a gem material. Its good yellow to yellow-green color makes it a suitable addition to the connoisseur's collection.

Cordierite: When cordierite, a magnesium aluminum silicate, occurs in gemstone quality, as it rarely does, it is known by one of several gem names. The most familiar of these to the jeweler and gemologist is "iolite," a word derived from the Greek, which alludes to its violet color. Perhaps the most outstanding characteristic of the gem is its very strong pleochroism. When viewed in the three different crystallographic directions, it is dark purplish-blue, yellow, and light blue. A strong bluish color is preferred for cut stones, but there is always a tendency for the color to be too intense. Well-colored iolite gems are rather rare. The hardness is somewhat above 7. Best known of its rare occurrences as a gemstone are the gravels of Ceylon and Burma, and scattered finds in Madagascar and India.

Danburite: As often happens, the locality at Danbury, Connecticut, which first yielded this species never produced any of gem qual-

130

*Opposite: Polished slab of labradorite
from Labrador oriented to show schiller (top),
flat oval cabochon of Arizona malachite
(bottom left), 60-carat faceted gem of sphalerite
from Franklin, New Jersey (right). Below:
Record-size, 24-carat stone of labradorite from
Oregon. Bottom left: Snuff box of lapis lazuli
from Afghanistan: Right: Gem crysocolla from Arizona.*

131

ity. Colorless, cutting-quality danburite has come from Mogok in Burma, Bungo in Japan, and San Luis Potosi in Mexico. A pale pink variety also occurs in Mexico. Good yellow stones have come from Burma and Madagascar. Danburite is a calcium borosilicate with a hardness of 7, and its gems are brilliant.

Diopside: This is a calcium magnesium silicate. Most of the gemstones it yields are green to brownish-green because of the presence of iron, which always replaces some of the magnesium. However, gems have also been cut from material that is colorless, bluish, or brown. Even good cat's-eye diopside is plentiful. Diopside has a hardness of only 5½ and a strong tendency to cleave, which limits its usefulness as a gem. Good green gemstones have been found in Piedmont, Italy; Zillertal, Tyrol; St. Lawrence County, New York; Ceylon, Madagascar, and Brazil. There is a chromium-bearing variety with a livelier green color which occurs at Kimberley, South Africa, and also at Outokumpu, Finland, and as cat's-eyes at Mogok, Burma.

Ekanite: Few living men have had the pleasure of finding new gemstones to bear their names. Mr. Ekanayake of Ceylon found the first sample of a new greenish gemstone in the Ceylon gravels in 1953. By 1961 the material was properly described and named in his honor and several new samples were found. This rare calcium thorium silicate has a hardness of almost 6½ but, unfortunately, is strongly radioactive. Asterism is common in the material with some gems showing good four-rayed stars.

Euclase: Well-formed natural crystals of euclase are rare enough to be worthy of a collector's attention. The mineral is a beryllium silicate with a hardness of 7½. Often colorless, crystals may also be completely transparent with strong enough shades of greenish-blue to be attractive gem material. The name alludes to its characteristic of cleaving rather easily, which reduces its value as a durable gem and turns the cutting operation into a gamble. Most cutting-quality euclase is found near Ouro Preto, Brazil, but it is known elsewhere.

Fluorite: Perhaps a discussion of fluorite —calcium fluoride—would be better placed in a text about industrial minerals. It is used in vast tonnages for smelting iron, manufacturing insecticides, and in the chemical industry generally. It is too soft (4), cleaves too easily, and has little to offer as gem material except its wide range of attractive colors. It may occur in several shades of pink, yellow, yellowish-brown, lavender, violet, blue, or green. Faceted cut gems are inexpensive and are made only for collectors. However, larger masses are sometimes carved. The carvings, though easily executed because of the softness of fluorite, tend to be very fragile because of the easy cleavage. Rather compact crystalline masses of a violet, purple and blackish fluorite, found at Derbyshire, England, are known as "Blue John." This material is a bit more durable and has been carved for centuries because of the attractiveness of the colors and patterns of polished masses.

Hambergite: It is questionable whether the natural specimens of this rare beryllium borate should ever have been cut into gems. But because collectors are always looking for something unique, some gems have been cut from transparent crystals that are free of distracting inclusions. The mineral is colorless, has a glassy luster and very little color dispersion. Cut glass would be just as attractive. However, a hardness of 7½ gives it some durability. The best cutting-quality crystals have been found in Madagascar.

Hematite: Like fluorite, this iron oxide is more at home among industrial minerals since it is the most important ore of iron. Now and then the mineral is found in pure, brilliant but opaque, black-looking crystals which lend themselves to certain gem purposes. These crystals have a hardness of 6½ and no tendency to cleave or fracture easily, so that they can be cut without too much trouble and will take a brilliant polish. Hematite has been used to make black beads which look like jet black pearl. Because it is so easily carved it has been commonly used to make cameos, intaglios, seal stones, and various kinds of ornaments. "Alaska diamonds" are really hematite.

Kornerupine: There is little chance that kornerupine, a rare magnesium aluminum iron borosilicate, will ever be a popular gemstone. All the material found so far in Greenland, Madagascar, Ceylon, and elsewhere is an odd shade of green or a greenish-brown. The typical gravel pebbles of the gemstone have a hardness of 6½ and show a strong dichroism. In one direction they are greenish to yellowish and in another they are brownish to reddish-brown.

Kyanite: Like andalusite and sillimanite, kyanite is an aluminum silicate. Its name refers to the normal blue color of this common species. Color in the very rare pieces of gem-cutting quality varies from blue to green. Kyanite is very difficult to cut. One problem is a marked variation in hardness—from 7 when measured in a direction across the crystal, to 5 when measured down its length. It has a strong tendency to cleave into flat blades. Beautiful sapphire-blue gems have been cut with difficulty from North Carolina kyanite.

Orthoclase: Earlier, this member of the feldspar family of minerals was described under moonstone, its most renowned variety. However, there is a beautiful, iron-bearing, transparent, golden yellow variety found at Itrongay, Madagascar. It's hardness of 6, rich color, and rarity make it popular with collectors, but it has little acceptance among commerical gems.

Painite: An extraordinary, mysterious gemstone, painite is a calcium borosilico-aluminate. The mystery arises from the fact that only a single crystal has ever been found in nature. It was recovered from gem gravels near the village of Ohngaing in Upper Burma by Mr. A. C. D. Pain in 1951. The deep-red mineral was determined to be a new species. However, its hardness (7½), color (red), density (4.01), and other characteristics make it difficult to distinguish from garnet, and it is quite possible that cut stones of painite already exist in gem collections masquerading as garnet. Eventually, the very lucky collector who is able to discover one will probably do so by examining its crystallography; painite is hexagonal and garnet is isometric.

Petalite: Again, this mineral—a lithium aluminum silicate—has been cut only to satisfy

the collector's whim. It has a hardness of no more than 6, cleaves under stress, and tends to be brittle. What is perhaps the largest cut gem for the species is in the collection of the Smithsonian Institution; it weighs 55 carats and looks like a well-cut piece of glass.

Phenacite: Because well-cut gems of phenacite look very much like the common rock crystal variety of quartz, they have never had much commercial value. The mineral is somewhat harder than quartz—almost 8 on Mohs' scale— but it is colorless, so that it has no other advantage except rarity, which counts heavily among gem collectors. Good cutting material has been found in some quantity in the Ural and Ilmen Mountains in Russia, and near San Miguel di Piracicaba, Minas Gerais, Brazil.

Pollucite: Rarity is almost the only claim that pollucite has to a position among gems. Only 6½ in hardness, it is colorless and glassy, and its refractive index and dispersion are too low to make the material attractive. It is a cesium aluminum silicate originally found at Elba, though the best-known gem-quality specimens are from Newry, Maine.

Proustite: It is difficult to imagine a gem with a hardness of only 2½—so soft it can be scratched with a fingernail. At the same time, this silver arsenic sulfide also has a disagreeable tendency to turn black over a period of time when exposed to light. Normally, as mined, it is a bright, rich red—a reason of sorts for the material being cut for gem collectors.

Pyrite: Pyrite is an iron sulfide, which when cut and mounted in jewelry is often mistakenly called marcasite. In spite of the name confusion, it does make rather attractive jew-

135

elry. Although pyrite is opaque it has a brassy color and bright metallic luster. There is no possibility that this will ever be a valuable gemstone material because it is quite brittle and is so common as to be found in dozens of places around the world.

Rhodizite: Found near Russia's Ural Mountains and in Madagascar, this rare borate (sodium potassium lithium aluminum beryllium borate) is one of the few borate minerals ever cut into gems. It lacks any distinctive color or fire. However, it cuts well, has a desirable hardness of 8, and has been cut into small but flawless gems for the collector.

Scapolite: Several mineralogical names have been applied to the scapolite group of mineral species: dipyre, wernerite, marialite, meionite, and mizzonite among them. The plethora of names arises because the composition of scapolite varies all the way from a sodium aluminum silicate (marialite) to a calcium aluminum silicate (meionite). Various other names are used for intermediate members of this series. Generally the hardness is 6 for all scapolites, of whatever composition. Luster is glassy and dispersion is low, but stones may be attractively colored in yellow, pink, or violet, as well as the usual white and colorless. Some stones have excellent chatoyancy, so that good scapolite cat's-eyes have been cut. When the chatoyancy is diffused and does not produce a sharp eye, cabochon-cut stones may strongly resemble moonstone. Burma, Madagascar, and Brazil have produced the best scapolite for faceted and cat's-eye gems, but the mineral is found in opaque, nongem masses in Canada, Mexico, and elsewhere.

Sillimanite: The mineral species sillimanite, an aluminum silicate, is chemically identical with andalusite and kyanite. Among gemologists it is better known by the name fibrolite, a name referring to the fibrous nature of most of the material. The gem-rich gravels of Mogok in Burma have produced some fine, transparent bluish stones suitable for faceting. Probably the finest of these cut gems is a 19.84-carat gem in the collection of the Geological Survey Museum in London. Gray-green chatoyant fibrolite has been found in Ceylon and fine, dark, almost black cat's-eyes have been recovered in South Carolina. Because of its fibrous structure, sillimanite varies from 6 to 7½ in hardness.

Sinhalite: The discovery of sinhalite in 1952 arose from an examination of an unusual sample of the gemstone peridot of a brown color rather than the usual strong green. It proved to be a new mineral species—a magnesium aluminum iron borate—and not peridot at all. It was named after Sinhala (Ceylon in Sanskrit), where it is found in the gem gravels. This brown or greenish-brown gemstone has few distinguishing features, and being rather rare, is little valued except among collectors.

Sphalerite: Seldom do we encounter gemstones that are more important as ores than they are as gems. This—like fluorite and hematite—is one of them. Sphalerite—zinc sulfide—is a major ore of zinc which is mined in tremendous tonnages in this country and elsewhere. Infrequently, a few transparent, gem-quality pieces are found showing fine yellowish-brown to orange-brown color instead of the usual dark brown to black shades. These are sometimes cut for the collector because, in addition to the attractive color, sphalerite has a very large color dispersion which gives it a fire more than three times that of diamond. The misfortune of this beautiful gem species is that, with a hardness between 3 and 4, it is too soft for almost any practical use.

Sphene: This is another gemstone with a strong fire—greater than that of diamond. At 5½ on the Mohs scale, its hardness is more than that of sphalerite, but not quite high enough to qualify it for general jewelry purposes. The golden-yellow cut gems of sphene, with their strong fire, are particularly attractive, but yellowish-brown, orange-brown, and green stones are also very fine as long as the color is not so intense that they become almost opaque. Switzerland, Burma, and Baja California are the best-known sources.

Taaffeite: No more than a half dozen gems of taaffeite, a rare beryllium magnesium aluminate, are known. All have been found by optical examination of cut gems thought to be spinel. Unlike spinel, some few of these stones were found to exhibit double refraction. Curiously, gem-quality taaffeite has never been found in place in nature. Identification is difficult except by observing the double refraction, because taaffeite has a hardness, density, and refractive index very much like those of spinel. Most of the gems, probably originating in Ceylon, are pale lavender pink, but at least one is a deeper, amethystine color. The gem is named after its discoverer, Count Taaffe, who found the first gem in 1945.

Tektite: There is considerable scientific controversy about the origin of tektites. These glass-like blobs, of varying dimensions up to perhaps three inches in diameter, often have strangely pitted, fissured, rounded, or otherwise oddly marked surfaces. The most exciting theory holds that they are extraterrestrial, having traveled through space until caught by the earth's gravity to come raining down on the surface. It is true that they are found only in selected spots on the globe, such as in Western Moravia and Bohemia near the Moldau River (Moldavites), and in parts of Australia (Australites). It is true, also, that although they resemble blobs of obsidian—a volcanic glass—their chemical compositions are unlike any known obsidian. Tektites provide transparent cut gems of a brown, brownish-green, or green color which are no more attractive or interesting than colored glass, except for the possibility of their celestial origin.

Willemite: Even for gem collectors it is sometimes difficult to acquire a cut gem of willemite. This zinc silicate is not rare at all, since it is mined by thousands of tons in the area of Franklin, New Jersey, and smaller quantities are found at occurrences around the world. However, crystals of gem-cutting quality are extremely rare and only a few have been discovered at the Franklin mines. The greenish-yellow gems are too soft (5½) and have a rather resinous luster; their interest is almost entirely for their rarity.

Zoisite: Any text or compendium on gems published more than two years ago will have little to say about the gem occurrences of zoisite, a hydrous calcium aluminum silicate. It was known then only as an attractive, pink, opaque, ornamental stone called thulite, or as

Opposite: 104-carat yellow fluorite
from Southern Illinois (top), rare 24-carat
step-cut axinite gem from Baja
California (bottom left), 8-carat kornerupine
from Madagascar (right). Below: Unusually
fine 28-carat gem of andalusite
from Brazil. Bottom: Very large 110-carat
sinhalite gem, a rare species from Ceylon.

a brilliant green, granular material from Tanzania containing bright red, but opaque, ruby crystals. Suddenly in 1968 and 1969, quantities of extremely attractive gem zoisite began to flow into the markets from prospects near Arusha, in Tanzania. This material is in flawless crystal fragments exhibiting excellent pleochroism. The predominant colors are bright sapphire blue and rich purple. When cut in the appropriate crystal direction, or when heat-treated, the rich color rivals that of fine blue sapphires. A hardness of only 6 and a slight tendency to fracture have not aroused any popular resistance to the introduction of this new gem because it is so very beautiful. Its origin in Tanzania quickly inspired the variety name of Tanzanite for the gem.

Some Ornamental Stones

Agalmatolite: A compact, soft material consisting of impure mixtures of one or several minerals. Most agalmatolite is composed of steatite, or soapstone, which is a hydrated magnesium silicate. It may also be pyrophyllite, which is a hydrous aluminum silicate. Whatever the soft, soapy-feeling material may be, it is gray, green, brown, black, white, yellowish, or often mixtures of several of these colors. The material is easily carved with a knife.

Alabaster: A white, very soft, and very fine-grained variety of a mineral called gypsum. Gypsum is hydrous calcium sulfate, is very common, and has a hardness of only 2. The softness of the mineral plus its uniform texture make it ideal for carving purposes. There is also a fine fibrous variety of gypsum known as satin spar because of its pearly sheen.

138

139

Amazonstone: The feldspar family of minerals has several species. One of them, microcline, is a rather common potassium aluminum silicate. Occasionally, the species is found in a beautiful, opaque, bright green—presumably the color of the Amazon River. Because it is sufficiently hard and durable, and cuts and polishes well, it is used for costume jewelry.

Anthracite: A very hard variety of coal which can be cut and polished easily. It once was popular for carving curios and souvenirs of the coal industry and coal-producing towns. Some of it was cut and polished for jewelry. It resembles the best jet, but takes a more brilliant polish and is harder than jet.

Aventurine: Most aventurine is a variety of quartz containing large numbers of inclusions of tiny scales of mica, hematite, or other platy material. These highly reflective plates give the material a flashing, spangled effect in bright light. Some feldspar produces the same effect with similar inclusions.

Californite: Although it resembles jade, this compact, somewhat translucent green mineral is actually an attractive variety of idocrase. Idocrase is a complex calcium aluminum silicate, and the variety name, of course, comes from its best-known occurrence in California.

Chrysocolla: Most of the gem material called chrysocolla circulating among amateur cutters and in the gem trade is not truly this very soft, hydrous copper silicate mineral. Rather, it is quartz which is colored bright blue to greenish-blue by varying amounts of impurities consisting of chrysocolla or other copper mineral species. The quartz contributes its hardness and ability to take a good polish. The lovely color is still that of true chrysocolla.

Crocidolite: Again, the available "crocidolite" is not actually this species, but is quartz which has either replaced or enclosed the original fibrous, asbestos-like crocidolite mass. Chemical changes in the original crocidolite color the mass in shades of yellow, brown, and blue. Because of its fibrous nature, crocidolite, when properly cut and polished, is the source of "tiger eye." Crocidolite figures, ash trays, bowls, bookends, and such, are quite attractive and popular.

Jet: As suggested earlier, jet is a black variety of lignite, or of brown coal. Like anthracite coal, it has a compact texture and polishes well, although it scratches and abrades so easily that constant repolishing is necessary.

Labradorite: This is a calcium sodium silicate, one of the series of closely related plagioclase feldspars which vary in relative amounts of calcium and sodium present. Pure labradorite has been found in flawless, transparent, colorless to yellowish to red pieces large enough to cut good, faceted gems. The best-known labradorite variety is the kind which exhibits flashes of spectral interference colors (described in Chapter 2). The intense and gorgeous play of many colors is caused by its thin-layered structure, or by thin, plate-shaped inclusions of other minerals, such as magnetite.

Lapis lazuli: The material most often cut and polished as lapis lazuli—literally, "blue rock"—is really a rock. It is composed of a highly variable mixture of the minerals lazurite (primarily a blue sodium aluminum silicate), pyrite (in brassy colored spots), and calcite (as white streaks and blotches through the blue). The best lapis lazuli is uniformly deep blue in color and consists almost totally of lazurite.

Malachite: The name malachite-green, when referring to pigments, evokes an image of a deep leaf-green color. This hydrated copper carbonate mineral has such a beautiful color—often distributed in interesting agate-like patterns—that it has remained universally popular for ornamental purposes for hundreds of years. Its extensive use in beautiful inlay work, such as that in Imperial Russian palaces, is legendary. Soft and subject to chemical attack, it is nevertheless prized above many other ornamental stones. Some malachite is quite beautiful when associated with bright blue azurite, another hydrated copper carbonate.

Mexican Onyx: Sometimes this soft, ornamental stone is known as cave onyx, or simply as onyx. It differs markedly from true onyx because it is not black and is much softer. It is a kind of banded, mottled, and clouded marble composed of calcite in various shades of green, brown, and off-white. Mexican craftsmen cut the soft material into a variety of objects for the curio and tourist trade.

Obsidian: This natural glass, resulting from rapid cooling of molten volcanic rock, occurs in colors ranging from green through brown to black. Common enough to be used for carving inexpensive figures and other curios, it is sometimes difficult to distinguish from man-made glasses.

Rhodochrosite: This is an unusual manganese mineral in that it has an intense, rosy-pink color instead of the usual black. It is an opaque, soft, easily carved and shaped manganese carbonate which is valued for general ornamental purposes. The best material comes from Argentina, some of it showing interesting agate-like banded patterns.

Rhodonite: Again, the prefix "rhodo" in this name means a pink mineral. It is a manganese silicate and, unlike rhodochrosite, is fairly hard at about 6. This makes it more difficult to fashion, but at the same time it is more durable. The fact that it is too opaque restricts its use to ornamental purposes, although it is cut and polished for inexpensive jewelry.

Serpentine: Because it may resemble jade so closely, serpentine is difficult to distinguish, even for experts. Generally, however, it is softer and has a lower specific gravity. There is a variety of serpentine called bowenite, with a hardness up to 6 and a specific gravity up to 2.9, which brings its characteristics very close to those of nephrite jade. Like jade, serpentine is variable in color and commonly found in black, brown, tan, white, and green masses. Capitalizing on its resemblance, serpentine is used to simulate jade carvings, although it is often beautiful enough to stand on its own merits. Verd antique is a white, calcite-veined variety often cut into large slabs for use as an ornamental decorative facing for buildings.

Variscite: With occurrences limited to a couple of small deposits in the United States, this hydrous aluminum phosphate has not been widely used as an ornamental stone. It cuts and polishes well, has a light but intense green color—often with interesting patterns of associated minerals—and is hard enough at 5 to be useful as cut stones for inexpensive jewelry.

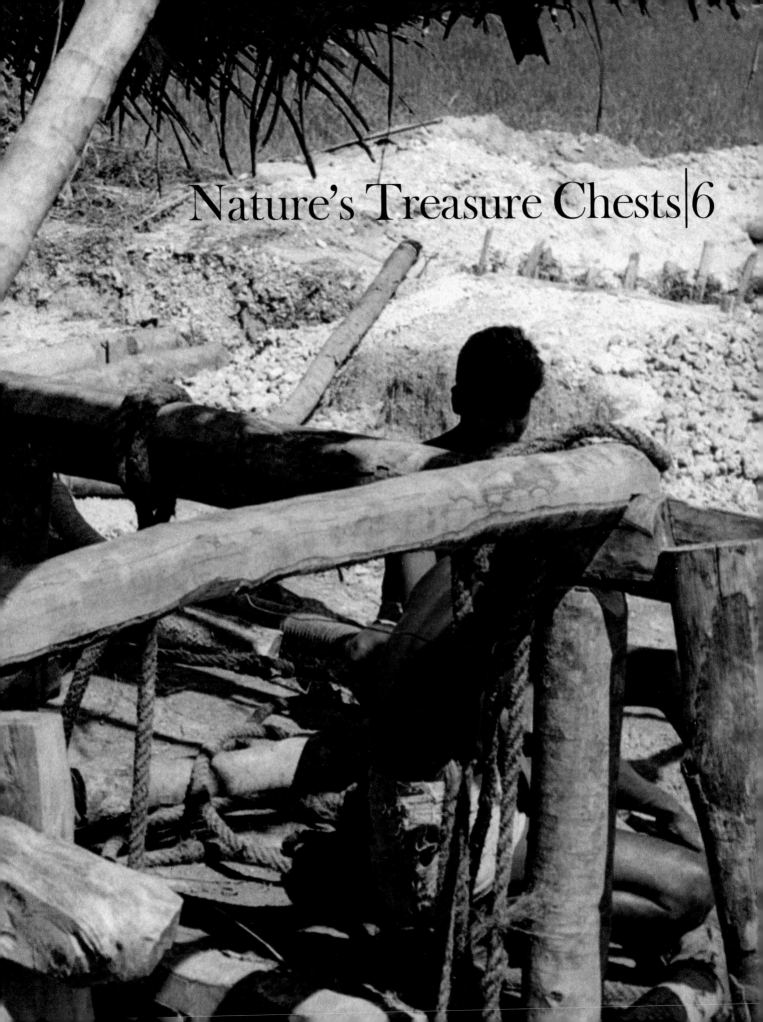

Nature's Treasure Chests|6

If the earth had been more free with its gemstones, perhaps man would treasure them less. This seems to be the case, with certain common species, such as quartz; some varieties of quartz rival the best gemstones for color and general beauty, yet they rank lower in the market place. It is the rarest gemstones that seem most highly prized. Often their rarity is extreme, some being found only in one or two accidental occurrences on the face of the earth. Gem books, describing remote regions of the world from which the great gems come, frequently dwell on the diamond mines of South Africa, the emerald mines of Colombia, the opal mines of Australia, and the seemingly inexhaustible gem gravels of Burma and Ceylon. Justifiably, these occurrences became famous and lore and legend accumulated around them. Some few of these unique deposits go through very long periods of sustained productivity and hold a place in the public consciousness. Others, reaching a peak of fame, suddenly are mined out and quickly forgotten. Six of the world's past and present sources of gems and gem materials are described here.

Strangely enough, one of the world's most productive gem material sources—one which cannot even properly be called a mine—is likely never to be exhausted. This is the group of cultured-pearl farms in Japan and elsewhere. Of course, these pearls are not totally natural gems since man initiates their growth. Indeed, pearls are not even minerals but are truly gems anyway, because of tradition and the fact that they have a place in the gem market and an impact on it.

The name Mikimoto is the one most often heard in connection with cultured pearls. A Japanese, Kokichi Mikimoto did more to develop the industry than any other individual. Yet there are paradoxes in the story of his success. Mikimoto appeared late on the scene—in the early 1900's—whereas the Chinese had practiced pearl culture as far back as the thirteenth century. Moreover, the process he originated for producing cultured pearls was soon superseded, even on the large farms of the pearl company he founded. Nevertheless, his name seems to persist as originator of the cultured-pearl method now in use.

The Chinese of old made crudely spherical pearls by introducing an irritating nucleus between the mantle and shell of the oyster. Their speciality was the production of small pearl Buddhas nucleated around small, hard wax images. Following these primitive efforts, there were several pearl-growing experiments by others which met with varying success. Even the great Swedish naturalist, Linnaeus, became interested and reported culturing pearls in 1761. He sold his process, which required implanting an irritant through a hole drilled in the shell, but it was never adopted commercially.

At the beginning of the 1900's, Tatsuhei Mise, an obscure Japanese carpenter, almost certainly understood how to induce formation of spherical pearls in the oyster species *Pinctada martensii*. In 1907 he applied for a patent on the process.

Five months later, one Tokichi Nishikawa applied for a patent on a substantially similar process. Nishikawa was a graduate in zoology from Tokyo University and had served with the Japanese Bureau of Fisheries until 1905, when he resigned to pursue research on pearls. While waiting for the patent office to act, he signed a process-sharing agreement with Mise.

Nishikawa's wife was Miyeko Mikimoto, eldest daughter of Kokichi Mikimoto. The husband and father-in-law never got along very well, and matters worsened when Mikimoto, seven years after Nishikawa, undertook to patent a somewhat different pearl-culture method he had developed.

Incredibly, Mikimoto's patent was granted first, on May 1, 1916. Nishikawa's came through on June 20—nine years late. And Mise's application was rejected, the authorities declaring, by a strange twist of reasoning, that it infringed on Nishikawa's, even though submitted five months earlier! Japanese patent law may have found for Nishikawa, but history will recognize Mise as the inventor of the method.

Eventually, father-in-law and son-in-law were reconciled, and Mikimoto's pearl farm began to use the more successful Nishikawa-Mise method. A strange sequence of events!

It must be remembered that externally a cultured pearl is indistinguishable from a totally natural pearl because it is made by the same animal, by the same process, and of the same materials. Only internally does it differ. Glass, plastic, and other spurious pearls lack almost all the fine gem characteristics of natural pearls. Cultured pearls, except for the negative psychological effect of man's interference in the process, have all of them. There are a number of pearl-producing mollusk species, but the most important for Japan's pearl indus-try is *Pinctada martensii.* When fully mature, the oyster is about 3 inches across, which makes it smaller than both *Pinctada maxima* and *Pinctada margaritifera,* the other pearl producers in this genus. Pearls grown by using *Pinctada martensii* are also smaller but much finer.

The cultured-pearl story starts with cultured oysters. Either naturally-grown "wild" oysters are collected or, more likely, a new crop is started from spat, the term for new-born oysters. Spawning oysters turn loose their eggs and sperm cells by the millions during July and August. The quickly fertilized eggs soon hatch out to produce enormous numbers of tiny, free-swimming animals that soon attach themselves firmly to rough surfaces where they will spend their lives. By setting out special cages and collecting devices the pearl farmer catches many thousands of these spat. The catches are left in place until about November when the young, half-inch oysters are removed and changed over to growing cages. Usually the cages are suspended from rafts, each supporting several thousand oysters, and are left until the following July. Now a year old, the oysters are about an inch in diameter. They are ready to be sown on the rough sea bottom or transferred to new growing cages. There they stay for two more years to reach their full size, vigor, and productivity. During the summer of their third birthday harvesting begins. The mollusks are collected and brought in to cleaning barges. There they are sorted and cleaned. All old and deformed shells are discarded; seaweed, attached animal life, and encrustations are removed. The cleaned oysters are

145

Sinhalese gem miner washing basket of gravel to free stones from coating of yellow mud. End product was few pieces of Ceylon sapphire at right.

again placed in growing cages for a few days of recovery before their ordeal of implantation.

A half hour or so before the operation, oysters are brought in and dumped on the wharf. Soon they begin to gape open and a bamboo wedge is inserted to keep them open. With special tools and a careful technique the operation proceeds quickly, so the oysters can be returned to the sea for recuperation before suffering excessive fatigue. The spherical nucleus around which the pearl will form is placed inside the tissues of the oyster, not between its body and the shell. The nucleus must be accompanied by a piece of graft tissue capable of producing pearl, because the normal internal tissue of an oyster cannot. Pearl nuclei are of many substances, one of the more popular being small spheres cut from mother-of-pearl. The graft tissue is prepared from an oyster's mantle, the thick flap of tissue which covers both sides of the body. From its surface, bearing against the shell, all new pearl to line the shell is produced. Cut into tiny pieces, one mantle will serve for many grafts. The pegged, or "keyed," oyster has a small incision made in its foot. A piece of graft tissue is inserted and the nucleus set on top of it, so that it is in contact with the pearl-producing cells of the implanted tissue. The foot tissue is smoothed over to close the incision, the retractor is removed to let the foot withdraw, and the key is removed so the shell can close. Now the oyster is ready to return to the sea. A very good operator can be successful surgeon to about forty oysters an hour.

With about 250 of its companions the oyster is suspended in a holding cage for a few weeks. Inspectors then remove all casualties, and the survivors, in their cages, go to permanent culture rafts. Such a raft may support sixty or so cages containing 3500 implanted oysters. Several times a year the oysters are cleaned of ocean life and debris and returned to clean cages. This continues over acres of rafts until the oysters are ready for harvest, three to six years later. The shells are then skillfully opened and the cultured pearls—plus any natural pearls formed as a bonus—are removed, washed, dried, and sorted by form and size. All this is done by hand labor to prevent damage to the pearls. Size sorting is done by a series of metal sieves of increasingly larger mesh. Those stopped by each sieve are kept together to be counted. The counter is a plastic plate with a handle. The plate has a hundred small, hemispherical depressions in it, so that when it is dipped into a bin of pearls it extracts exactly one hundred at a dip.

Next, the pearls are color- and quality-graded and readied to be strung, mounted in jewelry, or marketed directly. Production, although coming to a standstill during World War II, now numbers many millions of pearls each year.

It has been estimated that *Pinctada martensii* will deposit only one-eightieth of an inch of nacre in a two-year period. During that time there are many things that can go wrong. Cold ocean currents, which occasionally eddy into the warmer oyster bays, can depress water temperatures to the critical level of 11 degrees centigrade at which oysters will die. Winter air temperatures also have been known to cool water to killing temperatures. It is possible, by

*Opposite: Gem cutting in Ceylon
is done by little-changed, centuries-old
methods (top & bottom right).
Bottom left: 958-carat chrysoberyl
cat's-eye from Ceylon, very likely the
world's largest. Below: Diamond
mining in Brazil, just before diamond
rush to South Africa in 1870.*

The Kimberly diamond mine,
South Africa, as it appeared
(opposite) in the late 1800's.
Diamond-cutting and -polishing
equipment (below) dates
from same period. Above:
Hand-tinted plate from
Les Pierres Précieuses by
J. Rambosson, 1870, shows
emerald, sapphire,
garnet, and other gemstones.

towing rafts to warmer waters, to circumvent the problem at least partially. Higher than normal rainfall may also cause danger by reducing the salt level below the oysters' tolerance. Of course, there is always the chance that a violent storm will tear the rafts apart and send culture cages to the bottom. Regular cleaning, and the oyster cages themselves, help to protect the confined, helpless creatures from their natural enemies: the eel, octopus, barnacle, and seaweed. In the past, one of the worst natural enemies has been the "red tide." This is a large-scale invasion of red plankton organisms in such numbers as to cause the sea to turn red. In such quantities the tiny organisms can cause large oyster losses. One obvious defense is to tow the oyster rafts to safer waters at the first sign of a red tide invasion.

If the oyster survives, however, the purpose of its life—and all this work—is fulfilled. It produces a pearl externally indistinguishable from one it could have made by itself.

While a pearl is a gem by special dispensation, emerald—with diamond and ruby—is clearly among the most precious gems. Without doubt, too, Colombia has produced through the centuries the overwhelming bulk of gem emerald. Its total output must be expressed in tonnages rather than carats. In a report to the government of Colombia in the early 1900's a statement is made that "in the Muzo emerald mine the Republic of Colombia has a magnificent property, which should conduce to the prosperity of the country for generations to come." The report was optimistic, since history has decreed otherwise. The Colombian emerald mines have never been a

financial asset to the country but seem only to have stimulated dishonesty, greed, bribery, and corruption in individuals, business, and government. Even today, illegal mining and substantial emerald production by outlaws at the Peñas Blancas Mine, twenty kilometers northeast of the Muzo Mine, threatens government operation of emerald mining and marketing through its bureau of "Colombian Emerald Enterprise." It has been estimated that only 10 percent of the country's emeralds are sold legally.

Experts seem to agree that there are enormous quantities of emerald in large deposits in Colombia, but there is very little agreement as to how it all got there. Since the rest of the world's gem beryl has been found in high-temperature pegmatite deposits, there has always been the feeling that somehow high temperatures were involved in this case. Contrariwise, almost every scrap of evidence places Colombian emerald in low-temperature sedimentary rocks of related elemental composition. A search for low-temperature beryl in other parts of the world might well provide evidence to support the low-temperature theory. At any rate, it appears that the emerald crystals have been formed in isolated veins and cavities by percolation of water solutions through the sedimentary rocks.

All Colombian emeralds are found in two large mining districts. The Chivor district is about seventy-five kilometers northeast of Bogota in very rugged, almost inaccessible country heavily covered with thick cloud-forest vegetation. The mine sits 2300 meters above sea level on a mountainside. This mine and the

153

Top: Hand sorting of crushed rock for emeralds at Cobra Mine, in South Africa's Transvaal. Right: Sketch of Colombian emerald mine operated by forced labor in 1870.

Gachala Mine—thirty kilometers closer to Bogota—are the most important in the district. The rocks of this district are generally gray to black shales and sandstones, and the emeralds occur in them in rather sparse veins with albite, quartz, and pyrite. Only one other major emerald district is known. It is the Muzo district, with its center about one hundred kilometers north of Bogota. Tropical rain forest climate—hot and humid—smothers the district. Here the Muzo, Coscuez, and Peñas Blancas Mines are actively producing emerald. There is a difference in the rocks. They are black calcium-rich shales containing considerable organic matter. The emerald-bearing veins are much more abundant and contain several interesting minerals, including crystals of calcite, quartz, pyrite, parisite, and codazzite. The two districts, although separated by great distances, have certain geological and mineralogical similarities. The rocks of both are badly deformed and fractured, indicating considerable earth movement. Also their gem emeralds are usually badly flawed internally and contain numerous inclusions. These internal imperfections are called "jardin" (garden) in the gem trade. Often the inclusions make it possible to distinguish between Muzo and Chivor gems. Muzo stones usually contain tiny bits of dark organic material, while those from Chivor carry inclusions of tiny pyrite crystals.

Mining methods in both the Muzo and Chivor districts are unique, but have been in use since before the arrival of the Spaniards in the sixteenth-century. There were attempts by the Spanish to drive adits, or tunnels, into the Muzo deposits; at Chivor, a decade or so ago, tunnel mining was started. However, older terracing methods have persisted. The soft rock is worked out by hand tools along horizontal terraces rising in giant steps up the face of the mountain. All rock is carefully searched for emeralds as it is removed until eventually the working terrace becomes clogged with small, broken-rock debris. On hills above, large reservoirs are built and filled with water led in from the nearest source. Periodically, the flood gates are opened and the water, guided along prepared channels, rushes down and sweeps away the waste rock.

The Chibcha Indians mined Chivor before the Spaniards came and the precious gems they traded made their way from the Inca cities of the Peruvian Andes to the Aztec court of Montezuma in Mexico. Modern commerce has spread more of the same green gems all over the world, but they are still the same gorgeous emeralds extracted from the same mines by the same methods. Their appeal and value have been equally long-lasting.

The long-term, high-productivity histories of gem mining districts, so well illustrated by the Colombian emerald mines, South African diamond mines, and others, are in sharp contrast to the fates of some gem occurrences. As an example, opal mining in eastern Slovakia came to an ignominious end with development of the Australian opal mines toward the end of the 1800's. The Slovakian opals occur near Czerwenitza in Czechoslovakia. When the territory was part of Hungary, the town was known as Vorosvagas. Even today opal from this source is better known as Hungarian opal.

155

Hungarian opal can be distinguished from others by its generally milky-white background, although some of it carries tints of yellow or even red. The stones are translucent to transparent, with a play of the usual precious opal colors in flashes. The material is attractive but it does not have the bold color contrasts and brilliant color flashes so notable in Australian opal. Most of the opal recovered from these deposits is common opal in its many varieties: milk opal, hyalite, wax opal, and hydrophane, all useless for gem purposes. There is rather good evidence that opal of fine quality was coming from the district workings by the fourteenth century.

The rock in which opal was found at Czerwenitza is a grayish-brown volcanic rock left by one of several ancient lava flows in this part of Slovakia. Its occurrence was rather spotty, requiring removal and detailed examination of large quantities of rock. Even so, any blasting necessary was done with a "coarse-grained powder," since it was felt that the more efficient dynamite caused extensive shattering of gem opal. At first, mining was a rather haphazard surface operation of small pits and open cuts. Eventually, in 1788, the entire area came under the control of the government in Vienna, which attempted to organize the mining. Some underground workings were begun at this time but, on the whole, the effort was a failure and the mines were abandoned for many years. Several attempts were made to find the best operational system. Even a sharing program was tried. It was reported that, in Empress Maria Theresa's time, arrangements were made for the peasants to dig on two days

for the state and the next for themselves. Desultory official inspection almost encouraged the best finds to be made on the third day. By the end of the 1820's the government had begun to lease the mines, but mining methods remained primitive and disorganized. In 1845 Johann Goldschmidt, a Viennese jeweler, leased the mines for a period which was to extend, for him and his family, over three decades until 1880. During this period an extensive network of tunnels and galleries was driven into the opal-bearing rock. The development work was expensive but very productive. Not only did the government profit from the results, but so did hundreds of peasants employed as miners, officials, gem cutters, and as skilled and unskilled laborers. Of course, the lessees also prospered. A son, Louis Goldschmidt, traveled extensively to introduce Hungarian opal to the world. He exhibited at the great international exhibits in Vienna (1873), Philadelphia (1876), and Paris (1878), and maintained contact with gem markets everywhere.

The oldest mining activity in the area was on the slopes of the little Simonka Mountain, part of the southeast extension of the Carpathians. By 1870 it was mined out and shut down. Attention was then transferred to the Libanka Mountain range in the same general area. Enormous workings and mine buildings were installed and precious opal continued to flow from the district. Estimates had been made that this mine would be worked out by 1915, at which time Hungarian opal would suddenly become a rarity. Extensive prospecting had failed to uncover other sources. The Libanka

156

Mine never had a chance to die of exhaustion. It was killed by overwhelming competition from superior opal coming from the upstart mines of Australia.

Some gem mines do actually die of old age, or at least they seem to earn the right to some dormancy after centuries of exploitation. Turquoise deposits of the southwestern United States, in particular those of the Cerrillos district, come to mind as examples. One of these, lying about twenty miles southwestward of Santa Fe, New Mexico, is Mount Chalchihuitl in typical desert country. The name for the mountain is rather appropriate because it is the Indian name for a green stone. Very likely the word was applied specifically to jade, but it may have been used to label green stones in general. Most turquoise is greenish, only the rarest and best being blue. On this mountain, for considerably more than one thousand years, a succession of cultures has been mining turquoise. Archeological evidence is abundant to support the fact. What is nearly incredible at this mine is the enormous tonnage of rock that has been moved out by the Indians. By the use of crude hand tools alone the entire mountain was mined and quarried so that, on the north side, all that remains is a huge pit about 200 feet across and as deep as 130 feet in places. Undoubtedly, this spot supplied much of the turquoise that found its way into the Indian cultures in places as far apart as Yucatan and Ontario; there has been no significant turquoise discovery south of Mount Chalchihuitl.

Well before the time of the arrival of Spanish missionaries the gem was widely distributed among the Indians. Friar Marcos de Niza, in a report made in 1539, told of his journey toward what is now Zuni, New Mexico. There were enough turquoise beads, nose and ear pendants, and other jewelry, visible in every village for him to feel compelled to mention the gem fourteen times. Dr. Kunz, the gemologist, tells of attending a religious celebration in 1890 held by the Indians of Santo Domingo pueblo. Representatives of the San Felipe, Navajo, Isleta, Acoma, Jicarilla, Apache, and other tribes were present. The chief religious processional relic was a four-foot-high wooden image of Saint Dominic dating back to 1692. Before it were presented the Indians' most treasured gifts—turquoise beads and pendants.

There are no records of the amount of turquoise mined at these diggings during Spanish times. It has been estimated, however unreliably, that as much as $9,000,000 worth was removed from the district between 1895 and 1900 alone. Production during this time was boosted by the successful operation of the Castilian and Tiffany Mines, as well as a few neighboring prospects on Turquoise Hill, about three miles away. This operation was conducted by the American Turquoise Company, of which Tiffany's was the major stockholder and customer.

There have been many tales told about the Mount Chalchihuitl turquoise mine, as might be expected of a place of such antiquity. When the old mines were explored again in the early 1900's, many old stone hammers, crude mining implements, and pottery utensils were found. One hammer, made of local rock, weighed about twenty pounds; it must have

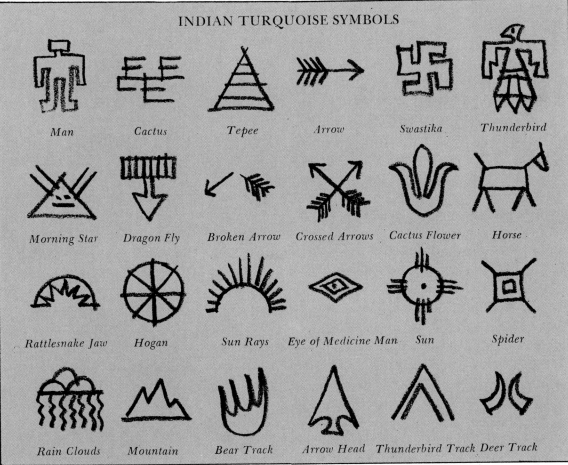

INDIAN TURQUOISE SYMBOLS

Man   Cactus   Tepee   Arrow   Swastika   Thunderbird

Morning Star   Dragon Fly   Broken Arrow   Crossed Arrows   Cactus Flower   Horse

Rattlesnake Jaw   Hogan   Sun Rays   Eye of Medicine Man   Sun   Spider

Rain Clouds   Mountain   Bear Track   Arrow Head   Thunderbird Track   Deer Track been wielded by a giant of a man. The presence of quantities of charcoal indicates that the ancient heat-and-quench method of mining was used. Fires were built against a rock face to heat it. The rock was then doused with water and the sudden cooling and contraction cracked it for relatively easy removal. (Water was an almost universal problem for turquoise mines, since the mineral occurs only in very arid regions. Not only was water needed for fire quenching but also for life support. Good water supply in the desert automatically means laborious transportation systems to put it where it is needed. Fortunately turquoise is never found at great depths; most mining can be done by open-cut or pit excavation. Nevertheless, between smoky fires, choking steam, desert heat, and stifling conditions, to say nothing of strenuous manual labor, it would seem that turquoise was hardly worth the trouble.) Evidently, disaster struck now and then at Chalchihuitl. Accounts vary, but there seems to have been a bad cave-in about 1680. One report claims that the entire mountain top collapsed, burying eighty Indian miners.

There still are quantities of turquoise to be recovered in the American southwest. Arizona, Nevada, and California still produce this gem mineral, but its quality is generally rather soft and porous. As a result, mining in these states is sporadic and generally inactive. Mount Chalchihuitl may perhaps come to life again some day, but at the moment it is dormant.

Usually, the gemstone occurrences that become famous and earn a place for themselves in gem literature are those that have produced commercially-important quantities of stones—

enough to establish a market. Some few, however, by historic accident, or because of the sheer beauty of the relatively few gemstones they have yielded, have also become notable. The tourmaline occurrence at Mount Mica, Maine, one of a series of New England tourmaline finds, is among these. It seems that two boys, Elijah Hamlin and his friend Ezekiel Holmes, discovered some fine tourmaline crystals lying loose in the soil in the autumn of 1820. There is little to distinguish the area now. A large hole, filled with water, is the only monument to the original find. Hamlin's son, Augustus Choate Hamlin, tells the story in his *History of Mount Mica*, published in 1895. The book itself is now a collector's treasure.

Once the find had been made, the boys were able subsequently to pick up thirty fine gem crystals of remarkable color and quality. Two of Hamlin's brothers opened the first large pocket in the pegmatite in 1822. From it came another lot of fine tourmaline crystals and mineral specimens. The secret was out and a rush started. Between 1822 and 1864 hundreds of pockets were opened. Most were barren of gem tourmaline, but many specimens were found which made their way, either as crystals or gems, into private collections. Careful, organized mining began around 1864. The gem pockets vary in size from just a few inches in diameter to several feet. There has been considerable weathering of the pegmatite, so that when a gem-bearing pocket is opened the tourmalines are found lying on the floor in debris consisting of loose mica scales, crumbs of feldspar, and clay.

Evidence of the past glory of this deposit

can be seen in two stones at the Smithsonian Institution: a marvelous, 57-carat, green gem cut from a crystal found in 1895, and 69¼-carat, bluish-green gem cut from a crystal found in 1893. Other fine gems and superb crystals can be seen in the collections of the American Museum of Natural History in New York and Mineralogical Museum of Harvard University at Cambridge, Massachusetts. Gem tourmaline from this and other New England occurrences is rarely seen on the commercial gem market. It is sometimes available for connoisseurs and gem collectors. Small-scale mining will perhaps be renewed periodically at Mount Mica and at similar Maine sites, but very likely as a labor of love rather than as a legitimate commercial venture.

It is too early yet to predict the future of a recently discovered new gemstone. From various museums and laboratories the news began to filter in, during 1967, that samples of an odd but beautiful gem variety of zoisite were trickling out of an unknown source in Tanzania. Simultaneously, and very quickly, the new material was identified and recognized in several laboratories as a significant discovery. Zoisite, a rather undistinguished-looking mineral, had been known since 1805. Even though tons of it had been recovered, none of it even suggested gem possibilities. True, there was a reddish to pinkish opaque variety, but it was known more as a seldom-seen, ornamental stone. When the Tanzanian type appeared, it quickly caught the fancy of gem collectors and merchants alike. Christened "Tanzanite" by Henry Platt, vice president and gem buyer for Tiffany and Company, the gem was popularized through extensive advertising and a demand for it established in the world's markets.

The gem is attractive because of its strong color. To the gemologist it is interesting because of its remarkable pleochroism. In the gem trade, cutters tend to classify the gem in three types: that which has strong pleochroism of dark blue, yellow-green, and red-purple; that with pleochroism of light blue, lavender, and colorless; and that which appears brownish because of pleochroism of green, purple, and light blue. The last type very conveniently turns to a rich sapphire-blue when heat-treated. The others need no heat treatment; their color is already strong and desirable.

What happens to Tanzanite depends on the supply. If sufficient quantities are found at regular intervals the gem will be a valuable addition to the jeweler's stock. If not, it will become a collector's item. At the moment so little is known of its source that no predictions can be made about the future supply. The gem was discovered in the upper part of the Umbe Valley, near Tanga in Tanzania, not far from the Kenya border. Mining is by rather crude pit and open-cut operations in a badly weathered metamorphic rock containing kyanite and other minerals. Practically no professional geological work has been done at the occurrence, since it is still in the first flush of its discovery and production. It will be interesting to watch the history of this gem discovery as it unfolds for us in our own time. Perhaps a thousand years hence some gem historian will note that for an incredible length of time quantities of gems have been pouring from the Tanzanite district—but perhaps not.

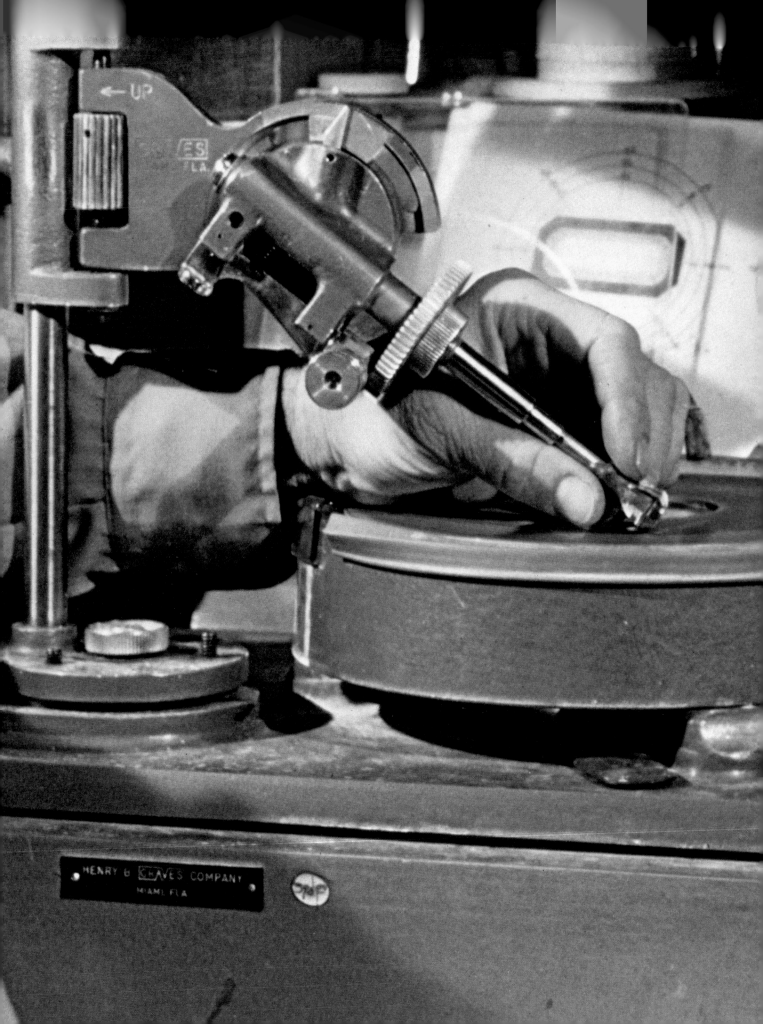

The Shaping of Stones | 7

For a gemstone to make the marvelous transition to a gem it must pass through considerable abuse at the hands of the lapidary. He is the man who selects a suitable technique with which to saw, grind, and shape each piece of rough material so that it achieves its maximum potential for beauty or salability. Nowadays it is assumed that most gemstones of any importance will be cut and polished by the lapidary into one of several standard forms commonly used for jewelry. This is a relatively new development in the history of gem preparation. Only gradually did cutting technology and tastes in jewelry develop to this point. Before the seventeenth century the general procedure was to grind and polish existing crystal faces or other flat surfaces to increase their transparency and reflecting ability. With many of the more colorful, opaque gemstone fragments it was common practice to round off the tops, sometimes producing rather oddly shaped gems. However, the art of the lapidary was advancing and by the seventeenth century lapidaries had learned how to take the fullest, most carefully calculated advantage of the potential of a piece of gem rough.

With certain technical cutting refinements, odd-form gems, now called "tumbled" stones, have seen a recent resurgence in popularity. Normally, they are cut from less valuable gemstones—such as the numerous kinds of agate—and are used for less expensive, costume jewelry. As might be expected, too, methods have been developed for their mass production. Very cleverly, the methods mimic the action by which nature tumbles and grinds pebbles in a stream bed. A hollow drum called a tumbler, which may vary from a hobby model the size of a coffee can to a large, heavy-duty commercial container, is charged with broken bits of gemstone plus a crushed or powdered abrasive and water. The sealed drum is then turned, for days on end, on a rack driven by an electric motor. After a series of abrasive changes, from coarse to fine, and after several weeks of continuous, night-and-day tumbling, the gems emerge bright, shiny, and colorful, and ready for key chains, bracelets, pendants, necklaces, and other kinds of inexpensive ornaments. Of course, more valuable material can be tumbled, too, but such pieces are usually reserved for individual treatment which will recover the greatest weight and therefore the greatest value from the stone.

In order to bring more order and less free form to these oddly shaped, tumbled stones, to draw attention to their beautiful colors and patterns rather than to their forms, and to prepare them to fit certain kinds of standard mountings, the cabochon cut was developed. Gem materials such as turquoise, amber, and jade are almost always cut this way for jewelry purposes. The word "cabochon" is derived from the Old French word *cabo* meaning head and referring to the rounded tops of the stones. Most jewelry cabochons are oval in outline with flat bottoms. There are simple cabochons with dome-shaped tops and flat bottoms, double cabochons with domes at top and bottom, and hollow cabochons with domed tops and scooped-out or concave bottoms. Very deep-colored garnets are traditionally cut as hollow cabochons to reduce their thickness and thus allow enough light through to show the color. Commercially, cabochons are often made in large quantities to fit jewelry mountings that are mass-produced in standard sizes.

In its essentials, the process of making cabochons is rather simple. First, the gemstone is sawed into slabs with a thin diamond-charged blade. The slabs approximate the finished stones in thickness. Using a template, the lapidary outlines the desired stone on the slab over a particularly choice pattern area. The good taste used in selecting the best pattern area may well be the most crucial element of a successful cabochon. The marking pencil is a sharpened rod of aluminum or brass. Once marked, the stone is trimmed out roughly by sawing and then by nibbling with pliers. Finally the stone is finished by grinding and polishing with ever finer abrasives.

The finest and most expensive gemstones are almost always cut as faceted stones. It is hard to believe that a process so basically simple as that for faceting gems—or for cutting any gem material—could have been a tightly controlled, trade secret for several hundred years, but it was. Faceting, like all gem cutting, consists of little more than attacking a gemstone, one way or another, with successively finer grades of abrasives.

It is likely that almost every hard material imaginable has been processed into service as a gem abrasive. Crushed sand, garnet, and corundum have proved to be hard enough for many abrasive purposes. For centuries the favorites for gem cutting had always been crushed corundum or emery, which is corundum mixed with magnetite or hematite. Both magnetite and hematite are readily available mineral species and ores of iron. Toward the end of the 1800's the introduction of a man-made abrasive, silicon carbide, brought about a revolution in gem-cutting technology. Silicon carbide—known commercially by names such as carborundum and crystolon—is harder (9½) than corundum (9). It is manufactured by fusing a mixture of 30 percent coke and 70 percent sand in an electric furnace. Under these conditions, the compound frequently is produced in masses of well-formed crystals with a beautiful, blue iridescence. For abrasive purposes the material is crushed and rather accurately separated and graded by the size of the grit particles. A typical coarse grit, called 100 grit, has particles averaging about $\frac{1}{170}$ of an inch in diameter. Such grit is ordinarily used for rough grinding, either as a loose powder or bonded on a grinding wheel. The particle size in 220 grit averages half the size of 100, and so on up to such small sizes as 1200 grit, which is used for the last fine grinding stage just before polishing a gem.

Powders for polishing must contain extremely small particles to avoid scratching an already fine-ground surface. The process of polishing is different from grinding because more is involved than just ordinary surface abrasion. While being rubbed rapidly with a polishing powder the surface of a gem may do one of two things. First, it may actually become hot enough for local melting to occur and for a very thin layer of melted glaze to form, giving the surface a high reflective luster. Second, if

165

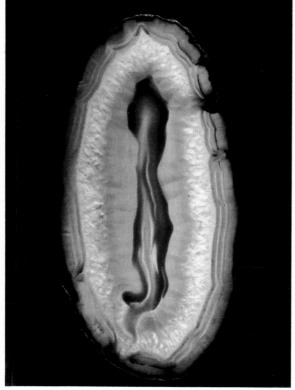

166

the melting point of the gem is too high for local fusion, the abrasion of the polishing powder may smooth out the tiny scratches left from the last grinding. It is more difficult to achieve a high polish when the fused layer does not form. This Beilby layer—named for G. Beilby, who discovered the effect in the early 1900's—flows over the stone and fills in the tiny scratches remaining. The effect is much like coating the stone with a clear plastic, except that the surface is much more reflective. In either case the results look about the same.

It is interesting that the polishing powder need not necessarily be harder than the gem it is polishing as long as it can develop enough frictional heat. Some polishing powders still used are quite ancient in the craft. Rouge, a powdered iron oxide, is effective but tends to leave red stains on the work, the hands, the machinery, and anything else with which it comes in contact. Green rouge, or chromic oxide, is not so effective a powder and also leaves a stain. Tripoli, a form of silica, is one of the more ancient polishing powders. It is fairly inexpensive but tends to be too coarse for present-day cutting requirements. Cerium oxide, tin oxide, and aluminum oxide are also widely used. Of these, aluminum oxide—and especially the artificially prepared Linde A—is generally the best polishing powder for all gems except chrysoberyl, sapphire, ruby, and diamond. It is manufactured with remarkable precision, so that no polishing problems result from variable powder size.

Costs may be a deterrent, but diamond, the hardest substance known, is the most superior grinding and polishing agent available. Carefully graded diamond particles are marketed according to the size mesh through which they will pass. Particles of 100 mesh will slip through a screen having 100 mesh holes to the inch. For heavy gem sawing, 50 to 100 mesh diamond is used. Fine sawing uses 120 to 250 mesh. Coarse grinding mesh is 400 to 600, and fine grinding is 800 to 1200. Polishing powders are prepared with mesh sizes from 3200 to 6400. These are excellent for polishing rubies, sapphires, and chrysoberyls. Of course, gem diamonds are so hard that they can be cut, ground, and polished only with diamond, which adds to the expense of these operations.

Armed with appropriate grinding abrasives, polishing powders, a wide array of potentially beautiful gemstones, and a highly developed metalworking craft, all that remained for the lapidary to develop was the cutting machinery itself. Elsewhere in this book (Chapter 10) some of the earlier, more primitive tools, devices, and procedures are described. In many parts of the Orient there has been almost no change in gem-cutting machinery since the days of Tavernier in the seventeenth century.

The gem-cutting craft was old even in his time. The first reference to a faceted stone appears as early as the 800's in a description of the gold cap of Duke Tradonico of Venice. The headpiece contained a sparkling ruby and a diamond with eight facets, as well as twenty-three cut emeralds. In the following centuries there were individual and carefully guarded innovations in cutting machinery and in the quality of the product. As might be expected, the flush of invention in Europe and the coming of the Industrial Revolution in the nineteenth century brought radical changes in lapidary machinery. And yet individual lapidaries and the cutting guilds managed to keep the nature of this new machinery, and the techniques for using it, a secret. Gradually, however, the secrecy was dispelled and gem cutting became the avocation of thousands of hobbyists.

In the United States before 1930, any amateur craftsman interested in gem cutting would have been hard pressed to find satisfactory abrasives or machinery, or, for that matter, any literature on the subject. The chances are that his achievements would be the result of trial and error, using homemade and substitute materials. By the 1940's, conditions had changed. Amateurs were inventing more new kinds of machinery and lapidary procedures than had been discovered in the entire history of gem cutting.

The old "mud saw," using a sludge of carborundum, grit, and water, was superseded by the diamond saw. Noisy, messy, and slow at best, it died without mourners, and today it is difficult to find one even as a nostalgic lapidarian antique.

The saw is basic to all lapidary operations. Perhaps it should be called a disk scratcher, since the word saw tends to evoke images of teeth, which the lapidary tool does not have. One can press a finger lightly against the edge of such a saw blade while it is in motion and not be cut, because the tiny diamond particles of its cutting edge are too fine to rip skin. The lapidary saw is made of a thin disk of bronze or cold-rolled steel set on an axle which can be turned at variable speeds by an electric motor with pulley arrangements. There are two types of disk, one with notches cut across its rim and diamond grit hammered into them, the other with diamond powder held firmly to the rim by sintering, or heating, with powdered metal. This powder-metallurgy process permits metallic bonding of the diamond grit at temperatures low enough so that the diamond is not destroyed. Saw blades are manufactured in various diameters and thicknesses. Slabbing saws are larger, with coarser grit. Trim saws use smaller, thinner, lighter blades and operate at higher speeds. For slitting very valuable gem materials there are paper-thin blades that rotate at high speeds. All gem saw blades are lubricated to keep them cool during cutting, so they don't warp or overheat the gems. In the early days an excellent lubricant for saws could be mixed at home with kerosene and motor oil. Later the major oil companies developed more satisfactory and less hazardous saw lubricants.

With all saws the gemstone is pressed firmly against the edge of the moving blade, either by hand or with a mechanical feeding device, until the diamond-edged blade literally scratches its way through. Because of its extreme hardness, it cuts quite rapidly. The gemstone rough is now sliced into appropriately sized and carefully selected pieces. It must next be ground to a shape approximating its final form. To strip off the first rough projections and unwanted material, grinders are used. Again, these are rotating disks. The disks, however, are made of silicon carbide which has been strongly bonded together by mixing with

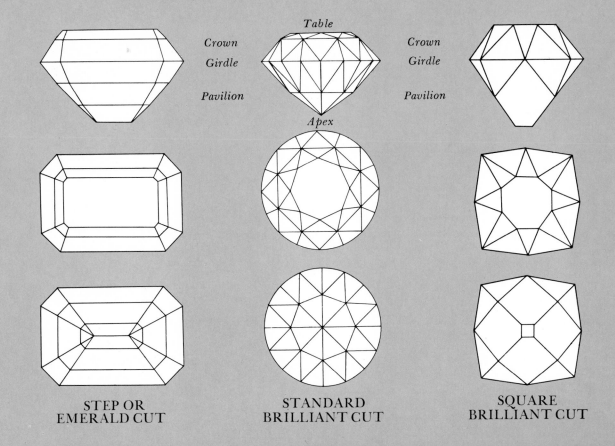

Crown

Table

Crown

Girdle

Girdle

Pavilion

Pavilion

Apex

**STEP OR
EMERALD CUT**

**STANDARD
BRILLIANT CUT**

**SQUARE
BRILLIANT CUT**

VARIOUS OLD-STYLE DIAMOND CUTS

Left: Egg-shaped, faceted, 7000-carat rock crystal from Brazil. Faceting and sapphire-studded stand are work of John Sinkarkas.
Below, left and right: Diamond cutters and polishers at work in early 1800's.
Bottom: Agate cutters of Idar-Oberstein.

clay and heating to a high temperature. These disks are also made in various sizes, such as 6-, 8-, and 10-inch diameters, with thickness increasing with the diameter from 1 to 1½ inches to improve physical strength at high rotation rates.

Grinding wheels are usually made of 100- and 220-grit abrasive and are kept water-cooled when in use. Whatever kind of grinding wheel is used, it is very difficult, if not dangerous, to hold the gem in the fingers, as is still done in primitive cutting shops. To eliminate most of the danger, and to make it easier to see the work in progress, the gem is usually mounted on a dop stick. Dop sticks are short pieces of dowel measuring from ¼ to ⅜ of an inch in diameter and 6 inches long. Actually, they are selected to fit the size of the gem and may even be as thick as a broom handle. The gem is heated a bit and fastened to the end of the dop stick with a special heated and softened dopping wax—much like sealing wax—molded up to support the stone during the cutting operation.

Once grinding is satisfactorily completed —the gem properly shaped and all large scratches and blemishes worked out—it is ready for sanding. The purpose of sanding is to remove gradually all of the obvious surface scratches and to perfect the shaping, but without major alterations. There are devotees of both wet and dry sanding but, except for the presence or absence of cooling water, the equipment and its operation are about the same. Sanding grits are usually 100 and 220, but are available up to 600 grit. Sanding is done with an abrasive bonded to the face of a

flat disk which fastens easily to the face of a rotating plate; this plate is sometimes attached to the end of the same axle that holds the grinding wheel. Or, the abrasive may be bonded to a continuous belt which fits an expanding drum rotating on the same axle. Provisions are made with most lapidary equipment to allow for rapid change of sanding-cloth disks or belts.

Now the crucial moment of polishing arrives, the moment which will tell whether the preliminary grinding work was properly done. There are all sorts of polishing devices, but rotating buffs of leather or felt are generally preferred. Water is applied to keep the buff damp and to help hold the abrasive polishing paste, which is smeared on as needed. Light pressure and movement of the stone over the buff continues until, suddenly, the lapidary realizes he no longer holds a common stone on the end of a stick but an object of beauty—a gem.

The Art of Faceting

The ultimate achievement for many in the lapidary craft is to facet a gem. For this the procedures follow the same general pattern as for cabochons, except that more precise and controlled methods of abrasion are needed. This means there must be specially modified equipment. Today's faceting equipment is so good that, armed with common sense and the ability to follow printed directions, anyone can facet a gem. The primary difference in faceting equipment is that all operations take place on disks, or "laps," which are mounted on a vertical shaft so that they lie flat and spin horizontally. A splash guard, consisting of a circu-

lar pan with a high lip, surrounds the disk to catch any abrasive sludge which may fly off the edge. The abrading surface is the upper face of the disk rather than its edge, and water is dripped onto it for cooling. Mounted over the lap is a faceting head. This is a vise-like apparatus which holds the dop stick in carefully controlled positions so that faces may be ground on the stone at exact angles to each other. Faceting heads differ but all have some means of indicating and controlling the angles precisely.

The traditional professional method of faceting employed a "jamb peg" rather than a faceting head. A jamb peg consists of a vertical rod holding a piece of hardwood in the shape of a pear or a flaring cone, and peppered with holes in more or less organized patterns. The pointed rear end of the longer dop stick is rested in one of these holes so that, as the gem-bearing end is lowered, the stone touches the lap at exactly the right angle to grind the face. It takes a remarkably skilled and experienced operator to know which hole to select and how much of a twirl to give the dop stick to position the next face. (It is no surprise to see many older cut gems with facets out of line or small facets inserted to close the gap left by mistakes in judgment.) And consider that, after all the facets are ground, it is necessary for the lapidary to retrace the whole course exactly in the polishing operation!

The procedures for faceting diamonds, because of their extreme hardness, differ somewhat from those for other gemstones. Only diamond grit can be used and great care must be taken not to sacrifice too much of either the

precious grit or the precious stone. In all cases the stone is first shaped, or preformed. Diamonds are often preformed by "bruting," or grinding into circular shape by being rapidly rotated on a lathe and ground against another diamond in a rod-like holder manipulated by the operator. Both diamonds are preformed in the process. Diamonds destined for other shapes are not bruted but preformed by grinding on a lap. Other kinds of gems are usually preformed on the 220-grit carborundum wheel. Once preformed, the remainder of the grinding and polishing of facets takes place on various kinds of laps made of steel, lead, type metal, cast iron, copper, or other metals. These metal laps may be fed with various loose grits, but a favorite lap is cold-rolled copper plate charged with diamond grit that has been pressed into it. Polishing laps may be made of lucite, tin, hardwood, or even linoleum and vinyl tile smeared with one polishing powder or another. Commercially, gems are faceted with fewer kinds of laps, generally using a loose diamond powder in olive oil as an abrasive. In recent years faceting laps have been introduced with resin-bonded diamond grit. They cut with great rapidity, so that their higher cost is offset by savings of time.

The finished product of this process has been referred to constantly in this book as a faceted stone. Such a stone can be defined best as a gem with a series of carefully placed, flat, reflecting faces. The choice of these faces and the angles they hold to each other are sometimes determined by the nature of the gem material, by its characteristic refractive index, and so forth. Sometimes personal taste alone

174

175

determines the placement of the facets.

In general, to describe a faceting style or cut, the stone is figuratively divided into sections. The flat top is called the table. Other facets surrounding the table but appearing above the middle of the stone are in the section called the crown. The middle itself is called the girdle. Below the girdle the facets lie in the pavilion section. If there is a tiny flat face at the bottom of the gem, opposite the table, it is called a culet. Owing to internal structure, exact angles for the placement of facets are different for every kind of gem material, but they are very frequently ignored even when known. There have been so many cutting styles developed that even the restrictions imposed by the shape and nature of the gem material itself are not overly limiting to the lapidary. Even so, popular taste has had some effect on restricting cutting styles to two basic forms: the standard brilliant cut and the step cut. A brilliant cut is round and its facets are triangular or kite-shaped. A step- or emerald-cut gem is rectangular in outline; its facets are sets of parallel rectangles. Modern jewelry design often calls for other cuts, such as the baguette, cut-corner triangle, kite, epaulette, keystone, and marquise. The skilled lapidary will know how to produce a scissors, jubilee, parasol, among many other cuts. Frequently, the kind of setting to be used determines the size and cut of the gem.

Gemstone Carving

The art and craft of the lapidary has never been limited to producing gems to be set in jewelry, however. Almost every conceivable lapidary technique has been used to process gem materials into objects not destined for personal adornment. One of these techniques—more an art than a craft—has been the creation of gemstone carvings. Untold tons of gemstone material are sacrificed to tourist-trade carvings of elephants, goldfish, roosters, gods, and Buddhas which are ground out by the indefatigable carvers of Hong Kong and elsewhere. These objects can hardly be called good examples of gemstone carving.

However, there are still very fine gem carvings being prepared in Asia. In Europe, the earlier carvings of Fabergé and the artist-craftsmen of Idar-Oberstein established a tradition for creative work. Now, with the present explosion of lapidary interest, an increasing number of artists have turned to gem carving. This has resulted in a steady stream of carvings and an increase in both the quantity and quality of the work.

Carving, like all other gem-cutting operations, is a matter of subtracting or removing unwanted material by sawing and grinding. The cutter must see the finished work in his imagination and remove everything that doesn't belong to it. To do this the stone carver uses any of the lapidary equipment already mentioned plus some special carving tools. Among these is likely to be a machine called a point carver. This is a motor-driven, rapidly rotating spindle on one end of which is a chuck for holding soft iron tools of various shapes. Grit and lubricant are fed to the stone as it is worked against the rotating tool point. Of course, carbide- or diamond-tipped tools can also be used at greater expense. Hole cutters, internal grinders, special machine rigs with flexible drive shafts, etc. are part of the arsenal of special devices.

There are size limits for cabochons and faceted stones. Although they are often cut in enormous sizes—cabochons several inches across and faceted stones of several thousand carats—such gems are reserved for show and museum display purposes. One flawless, faceted stone of yellow topaz from Brazil, in the Smithsonian Institution collection, weighs 7725 carats, or more than three and a half pounds. The carver may prefer to work on small pieces, but the upper size limit is set only by the weight he can handle alone or with the help of hoisting machinery. Stone carvings several feet tall are not uncommon. Of course, the value and size of available pieces of gemstone rough often limit the size of gemstone carvings.

Carvings may vary greatly in complexity as well as size. Simple surface inscriptions or raised designs on flat pieces require craftsmanship. Cameos or gem portraits are much more difficult because the finished subjects must be recognizable. Good three-dimensional, gemstone sculptures are most difficult of all. They require the same artistic ability as for any kind of sculpting, as well as considerable familiarity with gemstone cutting techniques. And each carving requires a good piece of stone.

The modern lapidary delights in finding new forms to produce, ancient forms to revive, and new techniques and machinery to make the process easier. Stone bookends, ash trays, table tops, candle holders, and innumerable kinds of objets d'art are coming from his shop.

Even the ancient craft of sphere cutting has been modernized. Spheres are better than almost any other form for displaying a wide variety of gem materials so that all their best colors and markings can be seen. Spheres are started by sawing out a block, sawing off its corners and grinding it into roughly spherical shape. Final grinding and polishing is done with two open-ended pieces of pipe with their inside edges machined to a forty-five degree angle. One piece is fastened to a vertical rotating axle and the other is held by hand. The rough sphere is set on the upright pipe and wet abrasive is fed to it while the other pipe is pressed and worked against its top. By proper manipulation of the upper pipe the sphere is kept rotating until perfectly ground. A change in abrasives will get it ready for polishing. This last polishing is done by covering the edges of the pipes with felt or canvas to which polishing paste is fed. For larger spheres larger pipes are used; the final product is about one-third larger than the diameter of the pipe. Although spheres are still preformed as always, there is now a machine on the market which takes the drudgery out of the final grinding and polishing.

And so, development of the art and technique of ornamental and gemstone cutting by amateurs and professionals proceeds. At its best it produces gems and other lapidary objects that are technically, if not artistically, much better than those made by the ancients. Happily, these objects can now be made in a fraction of the time, with greater ease and at far less expense for the delight of many, rather than the enjoyment of the wealthy few.

The Fashioning of Jewelry | 8

One would be hard pressed to justify jewelry as anything other than a luxury. Of course, jewelry is an extravagance but so are many things in any stable, wealthy society that has reached an advanced state of culture. Flowers, costume, furniture, paintings, automobiles, and other coveted objects all belong, at least partially, in the luxury category. Jewelry at times has been appreciated as a necessity, too. There have been periods in history when jewelry was as necessary as food, clothing, and shelter. One was required to wear some jeweled ornament as a badge of rank, to advertise economic solvency, or to cure, or prevent, a threatening disease or disaster. In contrast, our present attitudes about jewelry seem based on ideas of rarity and beauty, the dictates of fashion, and a happy state of mind, rather than on feelings of necessity. The old French words *jouel*, or *joiel*, which have given us our word "joy," may also be the origin of "jewel"; It is interesting also to see the similarity of the word *jouer*—to play. What could be more appropriate to our own concept of jewelry than to think of a jewel as a pleasurable plaything?

In a narrow sense jewelry can be defined as gems or gemstones mounted in noble metals to enhance their beauty and to make them practical for personal adornment. Increasingly, it has become necessary to consider as jewelry certain noble metal fabrications which contain no gems at all. Such an idea predates us by several thousand years. Further, the metals may be enameled or set with imitation or synthetic stones. Indeed, the metals need not be precious. At times, copper, stainless steel, brass, and plastics of various kinds elbow their way into this elite company. Even so, the prime pieces of modern jewelry, as in several centuries past, use real gemstones—the more precious the better—and the noble metals.

If a list was made of favorable characteristics for a suitable jewelry metal it would certainly include requirements for beauty as well as high resistance to wear and chemical attack. Such a metal would have to be workable into complex forms. It should be generally available, but not so common as to be uninteresting. Surprisingly few metals meet all these qualifications to any degree. Gold places first; silver and platinum just about complete the list. These three metals and their alloys—mixtures with baser metals—are exclusively the metals of quality jewelry. Relatively minor amounts of the bright and durable metals palladium, ruthenium, and rhodium are also pressed into service with the noble metals. An indication of the difference in attitude toward these noble metals is the fact that they are customarily not even weighed by the same system. Gold, platinum, and silver are measured by troy weight, while copper and others are scaled with standard avoirdupois weights. A troy ounce is about 10 percent heavier than an avoirdupois ounce and there are only twelve ounces in a troy pound. For everyday use we are much more accustomed to the sixteen lighter ounces in an avoirdupois pound.

Gold is so inactive chemically that quantities of it are taken directly from nature as a nearly pure yellow metal. Even in its natural deposits it has been able to resist combination with other elements. Most gold is stored in national treasuries in the form of bars for use in the settlement of international trade and credit balances. However, increasing amounts of new and reprocessed gold are bought each year for use in the jewelry industry. Even when alloyed with other metals, gold retains its remarkable malleability and ductility. It is so malleable that one ounce can be beaten or pressed thin enough to cover an area thirteen feet square. Its ductility is so great that the metal can be drawn into wires as fine as $\frac{1}{2000}$ of an inch in diameter. These extremes of workability are far beyond the needs of the jeweler, but assure that gold can meet any demands made of it.

Since the pure metal is too soft to stand up under the punishment it gets in everyday wear, it is usually alloyed with other metals, such as silver, copper, nickel, and zinc. These can be chosen to give the gold its popular tints of yellow, white, green, and red while they increase its hardness. The quantity of gold in an alloy is indicated by the term "karat." Pure gold is 24 karat. Fourteen-karat gold, for instance, contains 14 parts gold and 10 parts of other metals to make a total of 24 parts. Sometimes gold is graded as to its "fineness," based on how many parts out of 1000 are gold. An alloy of 85 percent gold and 15 percent copper would be 850 fine. Yellow gold alloys contain silver, copper, and zinc. White gold has nickel instead of silver. Red-gold tints are produced by using more copper and less silver than in yellow gold. Green-gold alloys require the reverse: more silver and less copper than

in the formula for yellow gold

Under United States' law the price of gold is set at $35 per "fine" or 24-karat ounce. This price is carefully controlled and is applicable to gold used for monetary transactions between countries. A free market in gold also exists in which the price fluctuates somewhat. Except in periods of heavy speculation or international monetary stress the free price hovers close to the controlled price. Because gold is expensive there is a large demand for "gold-filled" and "gold-plated" objects, which cost less but maintain the appearance and wear-resistance of gold. The two terms—filled and plated—are somewhat confusing to the jewelry buying public. In simplest terms, the difference lies in the thickness of the coating of gold applied to a base of some other metal, such as copper. Plating can apply a layer of gold as thin as $\frac{1}{100,000}$ of an inch. Because this layer is so thin, the jewelry item will have the appearance of gold, but it will not survive much abrasion or hard wear. A gold-filled object has a much thicker coating. It may bear a notation such as 1/10—14K. This means that the surface layer is 14-karat gold and that it makes up one-tenth of the total weight of the metal. This is a much more durable coat, suitable for most jewelry work.

Silver is a very popular jeweler's metal for almost all the same reasons. Its beautiful, bright, white color, its high malleability and ductility, and its resistance to physical damage are very desirable characteristics. Unfortunately, sulfur compounds in the air tend to tarnish and blacken the metal, so that periodic cleaning is required unless an antique look is

Below: Gold repoussé breastplate over
1000 years old from excavations in
Panama. Opposite: Gold and quartz nugget
from Virginia. Bottom left: 17-ounce
platinum nugget from Colombia.
Right: Roman coins of 2nd century A.D.

183

preferred. (Silver cleaning is not a difficult chore these days, since there are a number of quick-cleaning products on the market.)

Like gold, pure silver is too soft for most jewelry uses. It is usually alloyed with other metals. The silver color is quite persistent even when fairly large percentages of other metals are added. The most popular of these alloys for table flatware and jewelry is sterling silver, a mixture of 92½ percent silver with 7½ percent copper, which increases its hardness considerably. Spring silver is sterling silver which has been rolled or drawn to reduce it to as little as one-tenth its original thickness. The process hardens the metal and gives it the spring needed for tie clips, money clips, and similar pieces. Coin silver is less fine, containing only 90 percent silver and the remainder copper. This is about the same alloy that Mexican and American Indian silversmiths use for their traditional jewelry. German silver, or nickel silver, is not silver at all. It is an alloy of 65 percent copper, 17 percent zinc, and 18 percent nickel. Hard and chemically resistant, it is used for imitation silver pieces and for inexpensive table flatware. White metal is used for cheap, imitation silver jewelry. It is an alloy of 90 percent tin, 9 percent antimony and 1 percent copper. Its very low melting point even makes it possible to use rubber molds for casting. The price of silver through the years seems to move steadily upward. Ten years ago it cost about $1 an ounce. Now it is well on its way toward $2 an ounce. The price is not controlled by law, so that rapidly increasing industrial need keeps driving it up. Even so, the metal is far less expensive than gold.

Platinum occupies the other end of the value scale. Its price fluctuates, too, but at its best it is at least twice as much as gold and may be five or six times higher. Platinum, like the other noble metals, is very ductile and malleable, and it can be cast. It can also be forged and welded because it has a very high melting point compared to gold and silver. Usually about 10 percent iridium, a rather rare metal, is alloyed with platinum to contribute hardness. Although platinum is more difficult to work, because of its very high melting point, its great strength, good color, and absolute resistance to chemical attack make it ideal for use with expensive gems, such as diamonds.

The art and craft of working metals to make jewelry has been practiced by every culture of man, at one time or another, for tens of centuries. Gold has the most ancient history of them all. Some of the gold jewelry created by craftsmen in ancient India, Egypt, and Assyria has preserved its original character in its descendants for over two thousand years. It is not surprising in modern times to see bracelets, earrings, and other ornaments with filigree work, quaint chainwork, and other features that are remarkably similar to those in pieces taken from Etruscan and Cypriot tombs. There was considerable similarity in some ancient metal jewelry even among widely separated people. Part of this is undoubtedly due to active trading, the migration and dispersal of various cultures, and the wanderings of the craftsmen themselves. Even hundreds of centuries ago there was a strong tendency for art to diffuse itself through the known world. Much of the similarity in the world's jewelry

Gold historically has
been used for ornaments
of both practical
and mystical value.
Ancient alligator pendant
was excavated at
Chiriqui, Panama. Weight
ornament was standard
measure for Ashanti
people of Africa.

185

Opposite: Sicilian amber necklace. Above: Gilt king and silver rook from 19th-century English chess set called "Field of the Cloth of Gold." Henry VIII is set with rubies and diamonds. Left: Detail from colonial American gravy boat showing superb work on silver "join or die" snake symbol.

*Modern craftsmanship in working
of precious metals is evident in gold
floral bouquet and base set on
petrified-wood slab by Simon Yavitz,
sculptor in metals and gems.*

is also due to the independent discovery and rediscovery of the relatively few effective techniques for processing precious metals. Because gold, for example, has certain working characteristics, there are only a limited number of fabrication processes that can be used to shape it. It can be beaten into very thin sheets, shaped by chasing and repoussé work, engraved, cast, formed into delicate filigree, twisted and bent into loops and chains, granulated, and soldered. All of these techniques have been used, forgotten, rediscovered, and used again with the ebb and flow of fashion changes and the rise and fall of the craft, along with the cultures that have nourished it.

Gold leaf has very little direct application in metal jewelry but has been used at times as a precious covering for wood or metal objects, to which it is applied by beating while hot or by use of various adhesives. Thicker sheet gold lends itself handily to jewelry work. Gold chasing and repoussé, using gold sheet, offer the craftsman considerable opportunity to express himself in a yielding medium. Chasing is done with small, rod-like hand tools—these days they are made of steel—which are used to press the desired design into the sheet. The sheet is held against a lead block or against a specially prepared pitch, and the shaping or ornamentation is produced by tapping the hard chasing tool against the gold with a light hammer. The process requires artistic ability, skill, and patience. Repoussé is the same process except that the work is done from the back of the object rather than by chasing against its face.

Engraving tools are similar to chasing tools but the variously shaped instruments are finished to a sharp edge. This is essential because, unlike chasing, the idea is to scribe, or cut, the metal rather than shape it. The tools are set in wood handles, since the engraving is done by hand. Designs are transferred to the metal and are carefully cut out by the engraver.

Chasing, repoussé, and engraving are methods giving relief to flat surfaces. Filigree, or wire work, adds still another dimension to goldsmithing. Gold can be formed easily into wire of various dimensions. The wire can even be patterned or beaded by pressing it in hard metal molds. Very thin wire, either molded or plain, can then be shaped, twisted, braided, and soldered into almost unlimited configurations. These forms, in turn, can be soldered to a gold sheet backing, or to each other if more open filigree designs are wanted. The general effect of filigree is one of delicacy.

Soldering is a process for using still another metal or alloy which has a lower melting point but which is compatible with the metals to be fastened. A "hard" solder is an alloy selected to make a very strong bond and which has a higher melting point. Such solders are used for silver, gold, and platinum work. "Soft" solders are low-melting-point tin-lead alloys. These are generally used by electricians and plumbers. The joint to be soldered, in any case, is cleaned thoroughly. A "flux," an easily melted substance such as borax, is applied. It melts and covers the joint, preventing oxidation as the required high heat is applied. An appropriate amount of solder is added to the joint and melted in to make a permanent union. There are numerous special methods

Opposite & below: Simon Yavitz sculptures, one in gold with jade berries on gypsum base, other with flowers and basket of 18-karat gold. Left: View of Eugenie Blue Diamond details setting of platinum and diamonds. Bottom left: Ring of rare and valuable black opal from Australia.

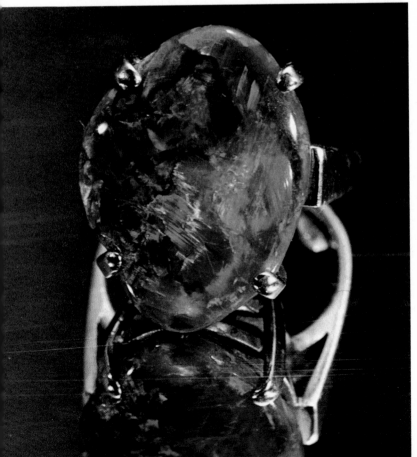

used to solder diverse kinds of objects and there are numerous solders to go with them. However, basically the process is this simple. Platinum, unlike gold and silver, has a very high melting point (3224°F.) and does not oxidize at high temperatures, so joints can actually be welded—melted together—rather than soldered. Some jewelry metal alloys contain a percentage of copper. After exposure to the heat of soldering, objects made of these alloys will blacken because of the formation of copper oxide. This "fire scale," as it is called, can be removed quickly by "pickling"—bathing in an acid solution which dissolves the black copper oxide and any excess, now-hardened flux.

There is considerable labor and time involved in progressing from design to finished jewelry piece. Casting, where feasible, is a method for reproducing a piece exactly, rapidly, and inexpensively. A mold is made of the model jewelry piece, using a special fine sand mixed with water and glycerine. The mold is divided, the model is removed, the mold is reassembled, and molten metal is poured in through an opening previously prepared for the purpose. When it cools, the new cast is taken out, excess metal removed, and finishing touches made. This kind of casting can only be done if the model is relatively simple and has no undercuts.

A far better casting process, especially for very accurate reproduction of delicate pieces, or those of some complexity, is the "lost-wax" method of centrifugal casting. The metal object to be reproduced is first copied exactly by making a rubber mold. In turn, the mold is used to make one or several copies in wax. The wax commonly used is similar to dental wax, with a melting point between 150 and 170 degrees Fahrenheit. Casting waxes are available in blocks or partially formed rings from which a model of a jewelry piece can be created directly by filing, sawing, and carving. The wax is much easier to work with than the metal itself. By either route—rubber mold or direct work in wax—the wax model is prepared, embedded in a plaster-of-Paris and silica mix to form a new, high-temperature mold. This is now heated in an oven to about 1300 degrees Fahrenheit. The mold hardens, and the wax melts and runs out. To insure complete penetration and casting of every detail when the molten metal is poured in, the mold and its contents are spun in a small centrifugal machine. This packs the metal into every fine detail of the mold made by the "lost wax." If carefully done, the finished casting requires very little filing or trimming when it is removed from the mold.

Through any combination of the techniques discussed here, by filing, buffing, drilling, etching, or in any other way of shaping the metal to his design, the craftsman finally produces a work of art. As with other artistic endeavors, each culture and each century has left its legacy of jewelry which tells us much of the state of the art and the taste of its devotees. Remarkably, the oldest gold ornaments we know were found in Spanish caves where they were left by their Paleolithic tenants. The famous wall paintings in these caves, and the crude, hammered gold amulets found in the cave sediments, date from about 40,000 B.C.

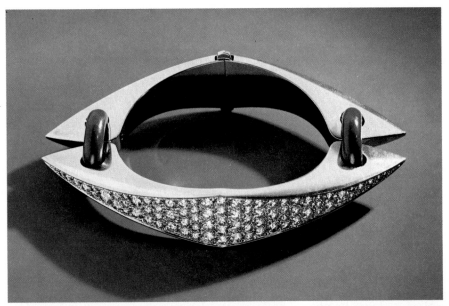

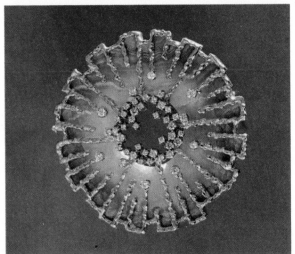

*Opposite: Brooch and
ring featuring Tanzanite
accented with diamonds
and emeralds; designed and
produced for Tiffany's,
New York. This page: Jewelry
designs entered in
1970 International Diamond
Awards Competitions.*

Certainly, gold was one of the first metals found by man. His copper and iron creations perished with use and age but the chemically inert gold persisted and so developed a place for itself as a special metal for special religious and secular purposes.

The ancestry of the jewelry we know can be traced in large part back to its origins in Mesopotamia. The great early civilizations developed along major river valleys. The Sumerian civilization was born in Mesopotamia between the Tigris and Euphrates Rivers. Egypt rose along the Nile. India ascended from the Indus River and, in the Far East, China developed on the banks of the Hwang. As these civilizations prospered, so did their goldsmiths. It seems certain that Sumerian goldsmiths had developed the art and craft to a high degree for their own use by 3000 B.C. Remarkable jewelry pieces have been recovered by archeologists during systematic excavations around Ur of the Chaldees, a great city which was the center of Sumerian culture at that time. In one grave—for Queen Shubad —there was an incredible treasure trove. The Queen herself had a kind of beaded coverlet made of gold, silver, lapis lazuli, carnelian, and other kinds of chalcedony. By her side were found hairpins of gold with lapis lazuli heads. She wore ornate head ornaments fashioned of gold rings, leaves, and flowers. Her ladies-in-waiting—buried with her along with all her other attendants—wore gold diadems and other jewelry.

Sumerians had such great enthusiasm for jewelry and such high artistic taste and skill with it that their influence on the jewelry of other cultures was strong. They handed their art down to the Assyrians and Babylonians. These in turn brought it to the Persians. The Scythians to the north and Hittites to the west were also impressed. The Scythians eventually introduced Sumerian jewelry culture to the distant Chinese, while the Persians took it to Greece, and the Phoenicians spread it throughout the Mediterranean world. None of these widespread cultures made significant improvements in either the art or craft through all these centuries, but at least they kept them alive and adapted them to their own forms.

The Egyptians and Indians had some appreciation of color in their jewelry. They added manufactured and natural stone and enameling to it, but by and large the best jewelry of most of the ancient world was wrought only in gold. However, by the time of Alexander the Great—about 350 B.C.—gemstones were more frequently found in Greek jewelry. Rock crystal, agate, sardonyx, carnelian, and even emerald added color. Interest in polychrome jewelry increased, stimulated by contact with Persia. The most riotous use of color in jewelry developed among Indian civilizations along the Indus River. An abundance of colored gemstones in this part of the world, the availability of pearls, and a knowledge of the techniques of colored enamel and glassy paste combined to make it possible.

By the time of the Romans, whose jewelry styles were based on neighboring Greek and Etruscan forms, the various pieces were primarily prized for the massive load of gems they carried. The art of the setting was considered relatively unimportant. When Byzantine jewelry finally evolved, its style was based on those of Greece and Rome but was far richer and more dazzling in color and design because of influences from the east. Tastes in jewelry design by the Middle Ages were governed largely by the ostentation of the display. Enormous and massive creations featuring girdles and large brooches of gems and metalwork were very popular. Later, during the Renaissance, there came along with social, religious, and political changes a revolution in jewelry design. Great artists, such as the Italian sculptor Benvenuto Cellini (1500-1571), the Flemish painter Peter Paul Rubens (1577-1640), and the Florentine sculptor Lorenzo Ghiberti (1378-1455), turned their attention to the design and production of jewelry. Their designs helped so much to enhance the beauty and popularity of gemstones and pearls that, by the late seventeenth century, the craft passed into the hands of the gem cutters and mounters. At the beginning of the eighteenth century there was a sudden surge of interest in highly intricate ornaments of over-enthusiastic design. Reaction was swift and tastes turned back again toward the early classic forms of jewelry. At this time, too, there was a great increase in the production of inexpensive jewelry pieces for the new, expanding, more affluent middle class.

Up to this point in time all jewelry was the end product of a small handcraft industry. Now, in the eighteenth century, factory-made ornaments began to appear. As with so many other everyday objects, quality suffered and the jewelry making art went into decline. The changes were not all negative. Jewelry wearing was finally brought to all the people by less expensive mass production; it was no longer considered a social privilege. True, certain kinds of very expensive jewelry are still indicators of economic privilege, but there is beautiful jewelry available for all. The new technology introduced another noble metal—platinum—for use in jewelry. Mechanization, too, allows for rapid changes in jewelry styles to match changes in styles of dress. Oddly enough, the popularization of jewelry has also brought about a strong revival of interest in new jewelry designs and in individually hand-crafted pieces. Each year, for example, awards are made for original designs entered in the International Diamond Awards competition. There are always some very imaginative pieces among the winners, which serve to encourage still more skilled artists and designers to apply themselves to this ancient pursuit.

At the moment jewelry making has at its command all the techniques for working gold, silver, platinum, and several other metals into almost any form imaginable. It has available the widest possible range of natural gems from all over the world, synthetic gems of many colors and characteristics, as well as glasses, plastics, and ceramics of every type. Best of all, there has been a surge of interest among artists and craftsmen of the highest stature to work in this medium. At the same time, a widespread interest in the hobby of gem cutting and jewelry making has spread among thousands of people who have the leisure and the means required. Altogether it seems to be the magic moment for important new things to develop in this very ancient art form.

197

As things now stand, royal crowns exceed by large numbers the current supply of reigning monarchs. Lord Twining in his book, *A History of the Crown Jewels of Europe*, supplies information about more than 600 crowns, 187 scepters, 98 orbs, and 116 ceremonial swords. A goodly number of these are still in existence and are carefully protected as part of the historic record of former kingdoms, as museum objects of artistic, aesthetic, and educational interest, or as regalia which will be required at future coronations.

The concept of the crown is an ancient one, its origins lost in, but traceable to, the dim past history of Oriental and Mesopotamian civilizations. Somehow, very early, a circular head ornament came to mean recognition of the sovereignty, dignity, achievement, or other special attainment of certain individuals by their fellow men. The use of a green-leaved wreath by the Greeks and Romans, for excellence in such diverse pursuits as poetry and athletics, is well known to us. In medieval times a kind of crown, which was really developed from a diadem, came into universal use for regal purposes. The diadem itself is ancient, having been originally a simple band of silk, wool, or other material tied about the head of a king to distinguish him from his lesser subjects. The cloth strip later became a band of gold. Still later in their development diadems became so embellished with gems and so ornate in worked precious metals that only royalty could afford them. Partly because of this direction of evolution in crowns, the ornament became very closely associated with the idea of monarchy and the name became an adjective for royal prerogatives. The meaning of crown colonies and crown lands is quite clear to us. Thus, "crown jewels" are the property of the reigning monarch and do not necessarily have any crowns among them.

The most recent imperial crown to be commissioned was made for an empress. Imagine the excitement and consternation in the firm of Van Cleef and Arpels of Paris when it was learned it had been chosen to make a crown for Farah Diba, who was to become, in 1967, the first empress of Iran. Respected as it may be, and quite accustomed to handling great fortunes in gems, it is still an event when a jewelry firm gets an order for a crown. Rising to the occasion, the designers and craftsmen created a crown in the best medieval tradition, making good use of the fabulous collection of gems in the Iranian royal treasury. It is a lovely fantasy in red, white, and green—Iran's colors. The most important gem in the crown is a superb carved emerald of 91.32 carats. To keep it company there are 35 other emeralds, 36 spinels and rubies, 105 large, natural, high-quality pearls up to an inch long, and 1469 diamonds for trim. Removal of these stones from the enormous collection was hardly noticeable. Iran has acquired these treasures in the past two hundred and fifty years and now uses them as backing for its currency, as well as for its coronations. Conversion of a few of them to objects such as a crown for a new empress only perpetuates a practice existing all through history. Crowns come and go; they are made, repaired, modified, taken apart, and designed anew, but the gems they contain are preserved and passed on from one crown to the next. At a time when the world was peppered with any number of kingdoms being joined, divided, conquered, and rebuilt, there was a steady demand for coronation regalia. Now the demand has fallen to almost nothing.

Surprisingly, several notable collections of former regal trappings have survived. Certain collections are still very much in use. Others, just as glorious, are not. The Crown Jewels of Imperial Russia, for example, housed in the Armory Museum of the Kremlin in Moscow, are a dazzling, rare, and valuable array. Making liberal use of diamonds, rubies, sapphires, and emeralds with precious gold and platinum, the Russian czars commissioned rich crowns, scepters, bracelets, diadems, buckles, earrings, and even entire bouquets of gems. It is true that in the hard period after the revolution, when Russia's economy was in ruin, quantities of art treasures, including valuable jewelry, fled abroad in exchange for goods and currency. There was even an auction of some of Russia's state jewels at Christie's Auction House in London in 1927. Nevertheless, many of the most important pieces were foresightedly kept intact in Russia. Peter the Great had unwittingly made preparations for this when, at the beginning of the eighteenth century, he ruled that all czarist treasures were state property. They were thus assembled in the capital at St. Petersburg. In World War I the gems were moved to the Kremlin for safety. There they were fortunate to survive both the war and the revolution which followed.

Perhaps the three best-known pieces in the Russian collection are the Grand Imperial Crown of Catherine the Great, the Imperial Scepter, and the Imperial Globe. Catherine's crown is heavily encrusted, by official Russian count, with 4936 diamonds weighing 2858 carats. Impressive as it is, the crown must be very uncomfortable to wear as compared with some of the older, sable-trimmed, Russian crowns. The Monomakh Cap, a Byzantine creation of the fourteenth century which was worn by Ivan IV at his coronation in 1547 and is now in the Kremlin Museum, is typical of these others. This doesn't suggest that the earlier crowns were less splendidly sprinkled with gems. The sable-trimmed Cap of Michael Feodorovitch (1613-1645), with a large topaz sitting astride a heavily jeweled arch, a large, pierced, octagonal sapphire, numerous smaller sapphires, several spinels, and a rather good Colombian emerald, certainly seems sumptuous enough. Catherine's crown is an attractive and tasteful but very ostentatious display of wealth. Some of the largest of the 4936 diamonds in it make up an arching band which rises from the circlet—front to back—and supports a huge, 398.7-carat ruby spinel at the top. The diamond-trimmed spinel in turn supports a small cross formed with four diamonds. The crown was made for the coronation of Catherine II in 1762.

Catherine, a German princess, had married Czar Peter of Russia, and at the first opportune moment had herself declared empress. Through the political skill and support of her favorite, Prince Grigori Orlov, and his brothers she was able to hold the throne. For

Chalice containing 1300 diamonds was given to Potemkin by Catherine the Great in 1791. State Crown of England has 317-carat Cullinan II diamond and Black Prince's Ruby (spinel). King's Scepter holds world's greatest diamond—530.2-carat Star of Africa.

Imperial Austrian Crown made by
order of Hapsburg Emperor Rudolph II
in 1604. Gold relief panels
portray coronations of Rudolph as Holy
Roman Emperor and Emperor of Austria.

various reasons Prince Grigori fell from favor. In an attempt to recoup his position and fortunes he gave Catherine a magnificent 199.8-carat diamond which he had acquired with this investment in mind. The gem, one of the great diamonds of the world, had supposedly been pried from one eye of an idol of Sri-Ranga in southern India. One wonders what may have become of the second eye. From hand to hand, at ever increasing cost, the gem eventually made its way to Europe and to the Russian court. The gift did not have the desired effect for Prince Orlov, but Catherine promptly ordered the extraordinary diamond to be set in the Imperial Scepter. Shaped like a half egg, the Chinese-rose-cut stone dominates the 20-inch gold and silver scepter, which is topped by a bejeweled metal-and-enamel double eagle design symbolizing Imperial Russia. It is interesting that the largest and most important of the Russian Crown diamonds is mounted in the scepter, like the greatest of the English Crown diamonds. Undoubtedly, it is a matter of practicality. They are both too big to fit anywhere else. The overwhelming Orlov Diamond is one of three which are of greatest importance in the Russian collection.

Another of these is the Shah Diamond, which was recovered about five hundred years ago in Central India. This stone weighs 88.7 carats, is bar-shaped—about three times as long as it is thick—and only partially cut. It bears three inscriptions which tell much of its history and has a shallow groove cut around one end. The groove, no doubt, was for fastening a silk or gold thread by which it was suspended before the throne of Aurangzeb, son of Shah

Jehan. The great French traveler-jeweler, Tavernier, reported seeing it there in 1665. The first of the inscriptions reads "Bourhan-Nizam-Shah-II 1000," which indicates it was owned by the ruler of the Province of Achmednager in India in 1591. The second reads "Son of Jehangir Shah—Jehan Shah, 1051." Since this date is our year 1641, the inscription strongly supports Tavernier's report of his court visit to New Delhi. The third inscription reflects the plundering of New Delhi in 1739, when the Persian conquerors made off with the Great Mogul's gems. It reads "Kadjar Fath Ali Shah," which was the title of the Shah of Persia in 1798. As a peace offering the diamond was presented to Czar Nicholas I in 1829 to atone for the assassination of the Russian ambassador to Persia.

The third best-known of the diamonds, but by no means the third largest in the collection, is the Russian Table Portrait Diamond. This is an irregularly shaped, thin, flat tablet which seems to be a cleavage piece from some larger stone. An Indian diamond, weighing about 25 carats, it is appropriately mounted in a gold-and-enamel, Indian-style bracelet. As for the other major diamonds, there are many listed in *Russia's Treasure of Diamonds and Precious Stones*, published in 1925, including cut stones of 57, 55, 47, and 40 carats.

The third most important symbol of imperial power in the Crown collection, the Imperial Globe, is designed much the same as other such globes. Made of gold, with a wide patterned wreath of diamonds about its equator, it has a similar band arching across its north pole. Surmounting all is an enormous,

rich-blue, diamond-trimmed sapphire which, in turn, supports a diamond cross.

Descriptions of the most important symbols of power—Grand Imperial Crown, Scepter, and Orb—only begin to illuminate the superb Russian collection, which is matched by few others save perhaps the Persian and English. The Crown Jewels of England stand alone for their overwhelming display of enormous diamonds of highest quality. Although there had been impressive and historic dia-monds in the English regalia for many years, the most spectacular additions came in the early 1900's. In 1905 Frederick Wells, super-intendent of the Premier Mine in South Africa, personally discovered the 3601-carat Cullinan Diamond. Named after Sir Thomas Cullinan, developer of the mine, it is the larg-est gem diamond ever discovered. In 1907 the Transvaal presented the stone to King Edward VII as a birthday gift. The Asscher Diamond Company of Amsterdam was commissioned to perform the cutting operation. Nine superb large gems resulted, along with a number of smaller ones. Of the nine the largest is the Great Star of Africa. It is drop shaped and weighs 530.2 carats. Alterations were made in the King's Royal Scepter—first used in 1661 by Charles II—to accommodate the stone. A succession of monarchs had altered and refurbished the scepter, but this last addition dominates and overshadows the remainder of the ancient three-foot staff. The ingenious clasp by which the stone is held can be opened to permit removal of the gem for occasions when it is worn as a pendant.

The second of the Cullinan stones weighs only 317.4 carats, but is still the second-largest cut diamond in the world. Room was made for this gem on the front of the headband of the Imperial State Crown, originally made for Queen Victoria in 1838. Large as it is, the diamond does not capture all the attention directed toward this magnificent head ornament. On the reverse side of the headband is the marvelous sapphire originally in the crown of Charles II. In a most prominent position on a diamond-studded ornament above the great diamond sits the brilliant red Black Prince's Ruby. This magnificent spinel, nearly 2 inches across, has belonged to English royalty since 1367, and is one of the most treasured gems of all. The Black Prince, son of King Edward III, received this gem as a gift from Don Pedro, king of Castile, and it was added to the Crown Jewels for the occasion of the coronation of his son Richard II in 1377. Just in this crown of Victoria's, Sir George Younghusband lists in his book, *The Crown Jewels of England*, 4

rubies, 11 emeralds, 16 sapphires, 277 pearls, 2783 diamonds—all in addition to the great stones.

Interestingly, the Imperial State Crown is not worn at the actual moment of coronation of English monarchs. They wear the much more symbolic Crown of St. Edward. In a way, Edward's crown is a counterfeit. There was a crown used for Edward the Confessor's coronation in 1043, and it is documented and described at subsequent intervals. However, the period of Protestant revolution and the turbulent reign of Charles I dealt heavy blows to both jewels and king. Charles was beheaded on January 30, 1649. On August 9 of the same year it was ordered that the royal treasure be turned over to the "trustees for the sale of the goods of the late king, who are to cause the same to be totally broken, and that they melt down all the gold and silver, and sell the jewels to the best advantage of the Commonwealth." St. Edward's Crown ceased to exist in this act of vindictive destruction. When Charles II was restored to his father's throne in 1660, it was necessary to create an entirely new set of coronation regalia. St. Edward's Crown was promptly reconstructed as close to its original form as possible. Thus, although its physical continuity in history is lost, the symbolic ties are maintained.

The third stone cut from the great Cullinan Diamond is drop-shaped like the largest of the group. Weighing 94.4 carats, it was available in 1911 and suitably ostentatious for the new crown to be made for the coronation of Queen Mary, consort of George V. As a matter of fact, there was an embarrassing supply of

great diamonds available. The 94.4-carat Cullinan was placed at the very peak of the crown, set in the cross which is supported in turn by a diamond-spangled orb. At the base of the crown in the front of the circlet is the brilliant-cut, 63.6-carat Cullinan IV. Each of these stones is easily removable for wearing as a pendant brooch. Amazingly, both gems had a difficult time holding their dominance in the crown. Set in prominent position at the front and just above the circlet was the legendary 108.93-carat Koh-i-noor Diamond.

Although this historic diamond had been presented to Queen Victoria in 1850 by the

207

East India Company, it was first worn in a crown by her daughter, Queen Alexandra. By this time it had already been reduced from its former 186-carat, high-domed Indian cut to a 108.9-carat oval brilliant. Earlier, the gem had come to the East India Company, which owned India, as booty at the end of the Sikh War in 1849. As certain as it is possible to be from ancient records, the diamond belonged in 1304 to the maharajah of a vast area of India now made up of Indore, Ghopal, and Gwalior. The gem left India and went to Persia with the Nadir Shah in 1739, when he invaded India, captured the Mogul, and made off with his treasures. When Nadir Shah later was murdered the gem was stolen and made its way through the hands of Afghan kings until eventually, through a refugee king, it became the property of Runjeet Singh in Lahore, India. There the British found it. In 1937, although still safe in England, the restless diamond was moved again from Queen Mary's Crown to be the glory of Queen Elizabeth's Crown in which it now occupies an almost identical position. The empty spot in Queen Mary's Crown was filled with the best available diamond—the 18.8-carat, heart-shaped Cullinan V. This stone, too, was mounted so it could be removed and worn as a brooch.

The Crown Jewels of Iran, already mentioned, are another collection of enormous value surviving relatively intact to our times. Remarkably, they were practically unknown to the world until this decade. In 1961 they were opened to public exhibition for the first time at the Bank Markazi in Teheran. The largest part of the treasure trove was gathered as spoils of war from the conquest of Delhi, India, by Nadir Quli Khan in 1739. In a history of this event, written in 1747, the Abbé de Claustre reported that along with great quantities of gold ingots there were thirteen thousand chests full of gold and silver coins and "there was also an inconceivable number of other chests filled with diamonds, pearls and other jewels." The collection now has undergone numerous additions and subtractions, but is substantially the same 240-year-old loot. Neither the Nadir Shah nor his successors considered the treasure as anything but a financial asset. They were little concerned with the marvelous jewelry that could be, and occasionally was, made from the available gems and precious metals. The bulk of the collection consists of diamonds, emeralds, rubies, sapphires, red spinels, pearls, and turquoises; there are also a few examples of other species of gems.

Already certain coronation pieces in the collection have achieved fame. Among them are the Pahlevi Crown, the Kiani Crown, and Empress Farah's Crown, mentioned earlier. The Nadir Throne in all its bejewelled glory claims as much attention as any of the crowns. Undoubtedly one of the most ornate and valuable thrones in the world, it is basically a single chair made of wood and standing 88 inches tall. Each piece in the assembly is covered with gold sheet, and is removable. In addition, thousands of gems stud the throne in intricate, symmetrical, typically Persian patterns. The throne was probably made for Fath-Ali Shah, the last of the Qajar dynasty, who began his reign in 1798. In the very center of the high, ornate chair back is mounted the most impor-

208

tant stone of all, an emerald weighing perhaps 225 carats. Arrayed about it are four other large emeralds, which may total as much as 550 carats. Spinels, rubies, emeralds, diamonds are there in abundance. Many blue sapphires dominate the panel at the foot of the throne, which pictures a reclining lion. The throne was used again for the coronation of Muhammad Reza Pahlevi Aryamihr as Shahanshah of Iran in 1967.

The Empress Farah Diba's Crown and the Empress herself attracted much of the attention at coronation time. However, the Shahanshah and his crown were the prime reason for the event. The Pahlevi Crown, as it is known, was made for Reza Shah, the father of the present Shah, who overthrew the Qajar dynasty and assumed power in 1925. Both he and his son have spurned the older Kiani Crown of the Qajar rulers. This newer crown is as resplendent in gold and gems as any monarch could wish. There are reputedly 3383 mounted diamonds that are used only for background patterns. To outline the peak and headband of the crown, there are 369 large natural pearls, many in matching sets. Centering a sunburst pattern of rays on the front is a cushion-brilliant diamond, weighing about 60 carats, of pale yellow color. There is also a 10-carat colorless, marquise-cut diamond set above the sunburst. The two major emeralds are truly magnificent stones, the largest, of about 100 carats, set at the rear. The other, 65 carats, is beautifully carved and mounted at the top of the crown. Two 20-carat blue sapphires flank the 100-carat emerald at the back.

The Kiani Crown is heavy and ungainly looking by comparison. It is built around a tall, round-topped hat made of stiff cloth. The gems are either sewn directly to the cloth or are mounted on metal plates which are sewn to the crown. Almost 1800 matched natural pearls were used as trim around every edge and to form a wide, solid band circling the center of the crown, interrupted only by other gems mounted in it. The largest emerald, of 80 carats, is mounted in the crown top ornament. The best diamond is a fine pink one of about 23 carats in the most prominent position, front and center. Hundreds of other smaller emeralds, diamonds, and rubies cover the crown. Other than the pink diamond, the most important gem is the very old and very beautiful 120-carat, dark-red spinel which is set at the top of the rounded dome of the crown. Engravings on the gem reportedly indicate that it was originally in the throne of Aurangzeb. This is the same throne mentioned earlier, before which Tavernier had seen the Shah Diamond suspended. Undoubtedly this spinel was part of the Indian treasure seized by Nadir Shah.

No other state collection of gems and jewelry pretends to contain the enormously valuable objects found in the "big three." Judging by historic, aesthetic, and artistic standards, however, there are state treasures just as important—so many in fact that several volumes of description would be needed to do them justice. Vienna, Copenhagen, Munich, Dresden, and many other European cities, large and small, have exciting treasure vaults. Almost every city or cathedral has some sort of gem treasure established by the largess of nobility or royalty.

Crown treasures of Denmark as shown at Rosenborg Palace in Copenhagen: Christian IV embossed-gold and enamel chalice. Queen's Crown dating from 1731. Diamond, emerald, and pearl jewelry used by Denmark's queens for 200 years.

Hard by the Botanical Gardens of Copenhagen across Øster Voldgade, lies the Kongens Have—the King's Garden—in which sits the lovely fairy-story palace of Rosenborg. Started in 1606 by King Christian IV, and substantially completed by 1633, it has been an extremely pleasant abode for a succession of kings named Christian and Frederick. King Christian VII was actually the last to live there, if only temporarily, while the English fleet was attacking in 1801. Now the building, retaining all its original beauty and charm, is a museum housing the "old Royal Collection of art and curiosities." Among these objects are the state jewels, a fine but modest collection which hasn't been needed since the coronation of Christian VIII in 1840. Though few, the crowns are exquisite. There is the Christian IV Crown, made in 1596, which is a marvelous, many-pointed, open and airy fantasy of gold and bright-colored enamels in which are set pearls and diamonds. It served for the coronation of two kings—Christian IV and his son Frederick III. King Christian V had his own crown made when he became the first absolute monarch of Denmark in 1671. All told, this new crown served four Christians and three Fredericks until Denmark's new constitution went into effect. The crown is designed somewhat in the English style, with eight arching diadems rising to the top, which bears a blue enameled orb and diamond cross. Smaller diamonds are used for trim, but the outstanding gems in the piece are two large ruby spinels and two large, deep-blue sapphires set front and back in the circlet.

The Queen's Crown was not made until 1731 and is generally of the same shape as the King's without the large gems set in. Displayed with Her Majesty's Crown is a collection of jewelry set in diamonds, emeralds, and pearls, which is still available for the use of the current Queen. The King may also use the jeweled stars and gold chains of the Orders of the Elephant and the Dannebrog. The gold objects traditionally used for the baptism of royal offspring are also pressed into service when needed. Actually, any of the Rosenborg collection should be available to the present royal family, since it remained their property even with the end of the absolute monarchy. There they sit and one can almost picture the moment when the crowns will be lifted from their cases, carried up the great spiral staircase to the enormous Knight's Hall, past the three large silver lions to the ancient ivory throne to be placed on the heads of the new sovereigns.

The collection of symbols of royalty in Denmark, even after the changes in government, suffered relatively few losses. By contrast, the collection accumulated to the south by the royal Hapsburgs has suffered grievous losses. And yet there is still such a treasure of the remnants to be seen in Vienna that, obviously, at its peak it must have been an overwhelming sight. The jewels and regalia of the House of Hapsburg were accumulated in part through historic accident and fortuitous marriages which brought the family various dignities and thrones. Responsible in part, too, was the strong urge of some members of the family to accumulate objects of art and craft for their collections. Archduke Ferdinand, regent of the Tyrol, who produced an illustrated catalog of his collection in 1602, was first of the great Hapsburg collectors. He acquired a large collection of historic arms and armor, an enormous collection of painted portraits, and many objects of the decorative arts, including works of Cellini. Ferdinand's nephew, Emperor Rudolph II, built a huge collection in Prague of paintings, sculptures, decorative objects in precious metals and gems, animals, plants, minerals, and quantities of other objects best referred to as curiosities. Part of his collection went to Vienna, but most of it was dispersed during the sacking of Prague by the Swedes in 1648.

One group of treasures acquired by the family through right of rule came from the Holy Roman Empire. Starting with the crowning of Charlemagne in A.D. 800, this empire managed to exist for centuries through a succession of difficulties and wars. In 1438 the succession passed to Albrecht II of the Hapsburgs and all the surviving imperial regalia of six hundred years of history came with it. By 1804 one of the Hapsburg line of Holy Roman Emperors, Franz II, had also assumed the title of Franz I, Emperor of Austria, just as Napoleon was assuming his title as Emperor of France. This brought together the kingdoms of Bohemia and Hungary, which he ruled, as well as all his own crown lands. Bohemia and Hungary had previously come to the family by marriage in 1515 and the death of the male hereditary heir to these two thrones in 1526. More was to come. Charles the Bold, last of his line, had died in the Battle of Nancy in 1477. His daughter Maria married Archduke Maximilian of the Hapsburgs. Thus, by mar-

riage again, the very important late medieval treasure belonging to the Burgundian dukes had come to the family. In 1736 some of the finest jewelry ornaments, liberally set with diamonds and other gems, were added to the collection when Franz Stephan of Lorraine married Maria Theresa of the Hapsburgs. The best of the diamonds among Franz Stephan's treasures was the Florentine. This 137.27-carat, light greenish-yellow diamond had once belonged to the Medici family. The great gem was eventually placed in the Hapsburg Crown and later formed part of a hat ornament displayed with the crown jewels. And so it went, one addition after another, with increasingly better accommodations becoming available to house them. By the 1730's all the treasures had been brought together in the Alte Burg, or Old Castle, in Vienna. What is left of them is there yet.

The First World War not only killed the empire and deposed the Hapsburgs, but also ruined Austria, leaving it powerless. In 1918 the royal family went into exile, along with all its personal share of the treasures, including the Florentine Diamond. Within two years Italy was insisting on the return of the regalia used by Napoleon as king of Italy. In 1932 Hungary demanded a return of regalia of the Order of St. Stephan and other Hungarian national treasures. Hitler took away the symbols of the Holy Roman Empire of the German nation to be returned to Nürnberg. Fortunately, Hitler's loot, but little else, was later returned. What remained after a series of such losses constitutes today's impressive display. Miraculously, the largest part of Vienna's remaining treasures survived the Second World War when the city was bombed extensively and several of its great public buildings were severely damaged.

In the *Schatzkammer* (treasure room) of the Alte Burg the treasure is divided into two sections, secular and ecclesiastical. Under Joseph II, son of Maria Theresa, the church and crown treasures had been jointly placed under the care of the castle chaplain. They are still together although in separate rooms of the treasure chamber. The major crowns are in the secular section. Most impressive of these is the Crown of the Holy Roman Empire. Often called Charlemagne's Crown, it is much more likely to be the crown made for the coronation of Otto the Great in Rome in 962. The single arch of the crown, symbolic of the ridge of a warrior's helmet, was replaced during the time of Emperor Konrad (1024-1039). Also, the cross dominating the front dates from as much as seventy years after Otto. There are no gems of great renown set in the gold plates of this octagonal crown. However, it is completely encrusted with gemstones and pearls, and with its four quaint plaques of enameled religious figures it is impressive in its antiquity and symbolism. Of course, the Austrian Imperial Crown, Orb, and Scepter are prime features of the display. The Crown, shaped like a modified bishop's mitre, has the traditional single arch running through the cleft. Richly decorated in gold relief scenes of the coronations of Emperor Rudolph II, Holy Roman Emperor and Emperor of Austria, it is also liberally trimmed with diamonds, rubies, and pearls. One very large sapphire tops the Crown, as is also true of

the Scepter and Orb. Interestingly enough, the shaft of the Scepter is made of a two-foot-long section of narwhal horn.

Chalices, crosses, swords, crowns, vestments, reliquaries, chains, ornaments—all heavily jeweled and worked in gold—make a dazzling show of the pomp and circumstance which has vanished from castle and country. The one memorable gemstone, the Florentine Diamond, did not survive its flight into exile. It was later reported stolen and its present whereabouts are unknown, although it is rumored to have been recut to unrecognizable size and form for resale.

Somehow one expects to find dazzling collections of treasures associated only with the royal houses of the former great imperial powers. And yet one of the largest, most exquisite, and most impressive collections belongs to a kingdom which no longer exists but is only a part of modern Germany. The domain may be gone but the *Schatzkammer* of the Kingdom of Bavaria still exists. Its treasures have survived and are displayed almost in their entirety in Munich. Crowns in this collection date back as far as the year A.D. 1000 and other objects are even older. The collection was not really organized until 1565. At that time Duke Albrecht V of Bavaria ordered that the treasures accumulated by his family, the House of Wittelsbach, would henceforth be established as a permanent treasure to be kept in the new palace in Munich. There they sat until the Second World War when the collection was removed to safer storage. This was fortunate, because the palace subsequently suffered severe bomb destruction. By 1958, however, the palace had

214

been rebuilt in faithful reproduction of the original and the treasure was placed on public display once more.

Empress Cunegunde's Crown is the oldest in the collection. She was the wife of Henry II, who ruled the Holy Roman Empire from A.D. 1002 to 1024. It is a simple circlet of five curved gold plates hinged together and bespangled with sapphires, amethysts, and topazes. Lovely as it is, this ancient crown is crudely fashioned with generally poor quality gems that are badly cut and set. It offers a startling contrast to the impressive and beautifully fashioned settings made in 1806 for the coronation of Max Joseph as king of Bavaria by order of Napoleon. The King's Crown is designed in traditional fashion, with the ornate circlet connected by eight gold diadems arching to a jewel-encrusted orb at the top, which supports a simple cross set with diamonds. Rubies, emeralds, pearls, and diamonds are present in profusion, but the largest stone is a deep blue sapphire set in the orb. Queen Caroline's Crown, made for the same event, follows a similar pattern but is considerably more ornate. Large pearls dominate the entire design, so that there seems to be little else. In reality, it too is well covered with rubies, sapphires, emeralds, and diamonds.

Spanning the years between Empress Cunegunde's Crown and those just described, the *Schatzkammer* exhibit contains a welter of crowns, diadems, and tiaras of every sort and description, dripping with gems set in precious metals, tracing the royal successions of a long line of related kings and queens. Since the crowns were so carefully preserved it is rea-sonable to suspect other treasures were also. There they are, including scepters, orbs, jeweled insignia of the Orders of St. Hubert and St. George and of the Golden Fleece, brooches, hat ornaments, necklaces, chains, bracelets, crosses, jeweled statues of St. George, portable altars, and even the jeweled and decorated private prayer book of King Charles, who reigned in A.D. 870. Thanks primarily to the original action of Duke Albrecht V, it is all there for us to see and admire.

France once had its share of glorious state jewels. Louis XIV enjoyed diamonds immensely. Tavernier has reported in some detail a thousand diamonds he sold to the French king. Among them was the great French Blue Diamond, which had originally weighed 112¼ carats when brought from India in 1642. The stone was recut to 67½ carats in a drop shape for His Majesty's use. It had been the custom to keep the coronation regalia of France in St. Denis Abbey near Paris. They were last used for the crowning of Louis XVI in Rheims in 1775. It was not too much later that the revolutionaries sacked the abbey and made off with the treasures. Fortunately, at least some of these were later recovered for safekeeping at the Garde Meuble, a treasure repository during the revolutionary period. Surprisingly, several great gems survived the first onslaught of the revolution. Among them was the renowned Regent or Pitt Diamond. Purchased in India by Thomas Pitt as a 410-carat piece of rough diamond, it was cut down to a beautiful brilliant of 140½ carats. This he sold to the Duke of Orleans, regent of France. Surviving the first political disturbances, it was swept off with all the gems and jewelry that could be salvaged to the Garde Meuble in the Tuileries. The story is recorded of a public exhibit of some of the booty. Secured by a steel clamp attached to a strong chain, the Regent Diamond was available for handling by selected representatives of the people. All seemed well and secure. Even the French Blue Diamond, set in a gorgeous jeweled insignia of the Order of the Golden Fleece, had reached the safety of this revolutionary treasury. Tragically, security arrangements were of good form but little substance. The interior doors were well sealed and guarded, so the thieves came barging in through an exterior window left unbarred. For three nights the thieves had their way and stripped the treasury. Eventually, the theft was noted and some objects recovered. The Regent Diamond came back, but the French Blue was gone forever. The Regent was found in the woodwork of a Paris attic.

In 1793 the French National Convention decreed that "all gold and silver, coined or otherwise, all diamonds, jewels, gold or silver lace, etc. that are found buried in the earth, or hidden in cellars, in walls, in garrets, under pavement, in hearths, in chimney flues, or in other places of concealment, be confiscated for the profit of the Republic. To anyone procuring the discovery of such objects a twentieth part of their value is to be paid." Undoubtedly, such measures brought gems and jewelry out of hiding, but little of it was added to the royal treasures. Most of it was converted to coinage or auctioned for cash. Political turbulence was to follow France for years—through revolution, the Napoleons, the restoration of the mon-

archy, and more revolution. Final disaster overtook the Crown Jewels in 1887 when they were offered for auction in London. A few pieces were held back, including the Regent, still to be seen in the Louvre today.

Reading about or visiting the marvelous collections of state gems and crown jewels around the world, one might easily get the false impression that all the most important gems, all the famous crowns and regal jewelry that survived were locked up in national treasuries and museums. To the contrary, many objects fit for a king are still in circulation. Some of these were once owned by royalty but many, though eminently qualified, have never been.

Perhaps the richest and most famous crown known that was never intended for the head of a king is the Crown of the Andes. This great gold crown is described as having "the circlet rising in eight points, pierced and embossed with intricately entwined acanthus scrolls and applied with clusters of table-cut emeralds in high carved settings simulating buds and flowers, eight similar stones set at intervals around the base below a narrow band of small emeralds. The four arches also pierced in an elaborate scroll design, similarly set, surmounted at their union by an orb and emerald-set cross, and supporting seventeen cabochon emerald drops which hang freely within the Crown." The story of this crown begins in the 1580's when a smallpox epidemic raged through Colombia. The city of Popayan, near the source of the Cauca River, was a prosperous cultural center in the path of the plague. As one, the people of the city prayed for deliverance from the death-dealing sickness and were

216

spared. In thanksgiving, the citizens donated gold and emeralds for a crown to be dedicated to the Virgin Mary. It was required that "the crown must exceed in beauty, in grandeur and in value the crown of any reigning monarch on earth, else it would not be a becoming gift to the Queen of Heaven." In 1599, in the Popayan cathedral, the crown was placed on a statue of the Virgin. In the early 1900's it was decided that the crown should be sold to build an urgently needed orphanage, hospital, and home for the aged. The fall of the Russian czars brought a halt to one possible sale being negotiated but finally in October, 1936, the sale was completed to an American syndicate. Still owned by the syndicate, it is now in the United States, but not on public exhibition.

There are extremely important gems, too, that are fit for a king but remain in private hands. Foremost among them is the Jubilee Diamond. It was a large, flattened piece when found in 1895, at the Jagersfontein Mine, in South Africa. In 1897, the year of Queen Victoria's diamond jubilee, the rough stone was cut to a 245⅓-carat, cushion-shaped brilliant of superb color, brilliance, and clarity. It is a truly remarkable diamond that also ranks as the third largest in the world. Soon after its exhibition in Paris in 1900 it was sold to Dorab Tata, an iron and steel tycoon of India. He had it until his death. In 1939 it was sold from his estate to Paul-Louis Weiller, a wealthy European patron of the arts. With his permission the great gem appeared in 1960 at the Smithsonian Institution for only the second exposure to public gaze in its history.

Important gems of this rank, now in private hands or held by gem merchants, tend to stay in circulation, sometimes with long periods of inactivity, until eventually they come to rest in a permanent museum collection. The odyssey of the Idol's Eye is typical. It was discovered in the Golconda diamond-mining area of India early in the seventeenth century, a diamond of excellent color and purity weighing 70.2 carats. Seized by the East India Company for a debt payment in 1607, it dropped from sight for almost exactly three hundred years. At that time it was known to be in a temple at Benghazi, Turkey. Stolen and later pawned in Paris, it was sold into private ownership in Spain. Still later it was acquired by a gem dealer in Europe from whom it was eventually purchased by Harry Winston of New York. (Mr. Winston has been quite accustomed to buying and selling important gems. Many of the largest and best stones in the world have passed through his salon. In 1958 he donated one of these, the 44½-carat, deep-steel-blue Hope Diamond, to the Smithsonian Institution.) After acquiring the Idol's Eye he sold it in 1947 to Mrs. Stanton of Denver. In 1962 it went on auction at the prestigious Parke-Bernet Galleries in New York, where the high bidder was Mr. Harry Levinson, a well-known Chicago jeweler. The gem has been in motion now for almost four hundred years. There is only a slim possibility that it will come to rest in some great state crown since there are so few crowns needed. More than likely it will be transferred from hand to hand among the new nobility of wealth or, like the Hope Diamond, enjoy permanent public exposure in a museum.

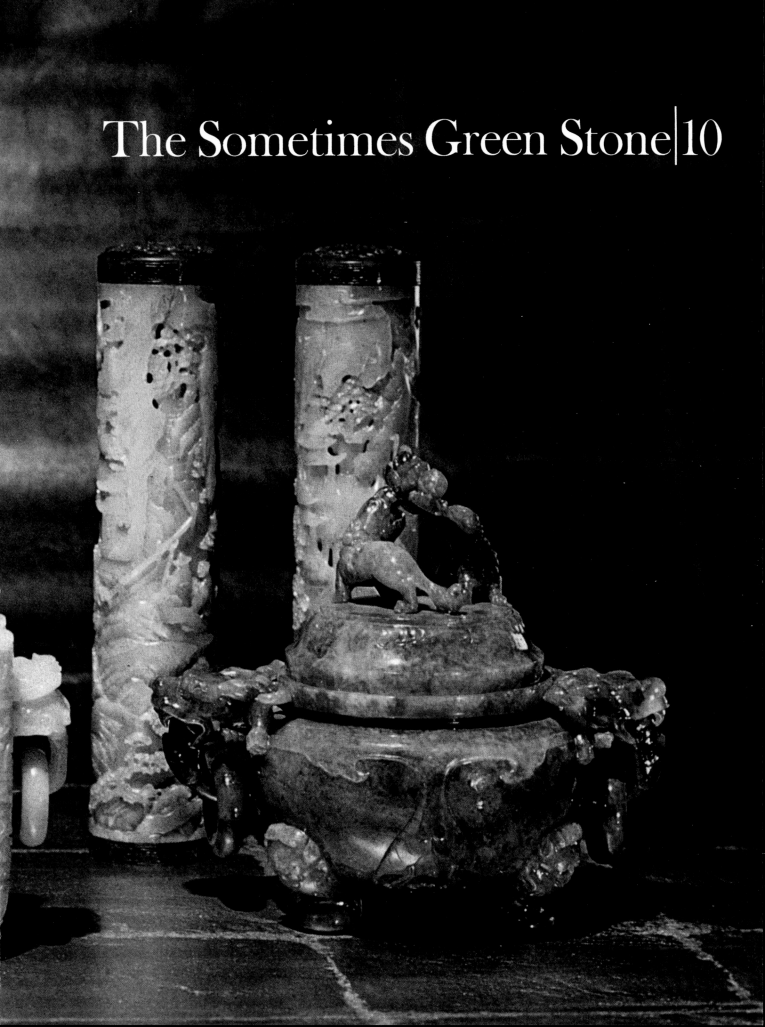

The Sometimes Green Stone|10

So much has been written about jade that it should be one of the best publicized and best understood of all the gemstones. On the contrary, most of what has been written as fact has been embedded in reams of speculation, guesswork, fantasy, fiction, and pure invention, so that jade remains a great mystery. Of course, a gemstone with a continuous popular history running through more than four thousand years is bound to accumulate a voluminous body of confusing lore. This is especially so with gem material that has been put to use for utilitarian, economic, political, and religious purposes along the way.

Jade has become what it is in history, legend, and gemology as a direct result of its inherent characteristics. Consistently, publications about jade dwell on its mystical and potent powers, and on the reverence with which it has been regarded by the Chinese. Actually, the gemstone needs none of this, because its characteristics alone can explain its exalted position. It is very interesting that even today an exquisite Chinese carving, having all the appropriate mystical symbolism, will suddenly lose its magic and market value when it is discovered to be executed in one of the very similar—almost indistinguishable—natural substitutes and not in true jade.

The facts about jade took gemologists a long time to assemble. A good part of the problem lay in the definition of jade. It was so broad that the name—or its equivalent—has been applied to several mineral substances with grossly different characteristics. The an-

cient Chinese employed the word *Yü* in a very broad sense for any good hard-stone carvings, including those of what we now recognize as "true jade." A more specific Chinese name, applied many centuries later in history to bright green jade, is *fei-ts'ui*. The name means kingfisher, and alludes to the bright green color of the feathers of one species of this bird. There was even a time when the name jasper, which we now use for a variety of quartz, served several kinds of jade.

Strangely enough, despite the close connection between China and jade, the names used for this material come from an opposite corner of the world. The name jade was established before the time Chinese jade and jade carvings were well known in Europe. There are no known references to jade, by any name, in the literature of Europe that predates the discovery of jade in America. It was the Spanish who introduced this "new" stone to Europe and gave it a name.

The Spanish conquistadores came to Mexico and Central America for various reasons, among them the urge for adventure and exploration. Very likely the booty they desired most was gold, and they found it. They also found *chalchihuitl*, which the natives prized even more highly than gold. This stone of great durability and bright green color was revered sufficiently by the Aztecs to be reserved primarily for royalty and high officialdom. Apparently, even in those times, it was rather rare and limited in its uses and distribution. Soon after the Spanish conquest the high art of American jade carving died. The existing carvings were plundered and dispersed. Native interest in the stone was lost along with the highly organized culture which valued it. Even the locations of the jade deposits were forgotten. There had been sufficient time, however, for the Spanish to import the stone. They identified it as very like the Chinese material then being brought in by Portuguese traders, and they adopted a strange mixture of Chinese and Aztec beliefs about it. One of these had to do with the miraculous curative power which jade could exercise on urinary disorders. They labeled it *"piedra de ijada,"* or stone of the loins. Given the similarity in Portuguese and Spanish languages, it is not surprising that the stone from both sources soon became known around the world by the corrupted name "jada." In French literature it was translated as *"pierre de l'éjade."* By repeated error it became *"le jade"* and in English literature it became simply "jade."

The material which thus came to be called jade was not the same as the mineral of like appearance that the Chinese had known for centuries. Chinese jade was what we now call *nephrite,* while the stone used by the ancient Americans has the modern name *jadeite.* Almost all of what is and was known as "true jade" is one of these two minerals. (Alas, the name of the Chinese species is not even of Chinese origin. When the Spanish for kidney stones was translated into Latin it became *lapis nephriticus,* which later became "nephrite.") Jadeite and nephrite really have very little in common. At times they may look very much alike and both can be very tough, durable minerals. Observed closely, the greens are not the same, even when both minerals are of good green color. The Spanish, for example, were able to generate some confusion between jadeite and emerald. This is possible because jadeite and emerald greens may sometimes approximate each other. Such confusion could not exist with the typical spinach green of nephrite. The differences in these hues are explainable because nephrite gets its green from the presence of iron in its composition, while jadeite's green—like that of emerald—arises from the presence of chromium.

Nephrite and jadeite both are mineral species arising from special metamorphic conditions. Seldom are these jades found in place. They usually occur as pebbles and boulders in present or ancient stream beds to which they have been carried from their original deposits by strong erosional forces. Nephrite is very often associated with other iron-bearing metamorphic rocks, such as hornblende schists and gneisses, and with its look-alike, serpentine. Jadeite, on the other hand, occurs with a variable suite of minerals that require high pressures and temperatures for their formation. Usually these minerals associated with jadeite include albite and quartz. As with nephrite, serpentine is often present. Jadeite and nephrite are almost never found in proximity with each other, attesting to the different sets of natural conditions under which they are formed.

Nephrite is a calcium magnesium silicate. It is a member of what is called the tremolite-actinolite series of minerals. Tremolite is a pure, white, calcium magnesium silicate, and actinolite is a green calcium magnesium iron silicate. There are any number of mineral sam-

ples whose composition falls somewhere between the two, nephrite among them. The tremolite-actinolite series itself is part of a larger family of mineral species known as amphiboles. Jadeite is not even closely enough related to nephrite to be part of the amphibole group. Nephrite differs from the other tremolite-actinolite minerals in that the typical fibrous crystals of the group are very compactly and tightly matted and meshed together. This produces an extremely tough, tufted structure. Its hardness is only 6½, which is not extreme among minerals. However, its toughness, due to the gross structure, is so great that very thin and delicate objects carved from it are not nearly so fragile as they look. It was an ideal material for primitive tools, which in some ways were more durable than the iron implements introduced later in Chinese culture. Nephrite is three times as dense as water, but its specific gravity, due to compositional and impurity differences among selected samples, varies from 2.90 to 3.01. This value, unfortunately, is too close to the density of its most troublesome substitutes, but serves admirably to distinguish it from jadeite. Sometimes nephrite is green. Certainly its best-known and more highly prized pieces are green. Shades of white, brown, yellow, and gray-to-black are just as natural to it. Even blue nephrite is known. Almost all these colors, as suggested earlier, are due to the presence of iron in its composition. Oddly enough, white nephrite has been found with considerable iron content, so that at least part of the reason for coloring in this kind of jade is still a mystery.

Jadeite is one of a group of minerals known as pyroxenes. Basically it is a sodium aluminum silicate. One variety of it found in Central America has an intermediate composition between jadeite and diopside—a calcium magnesium silicate; it is usually referred to as diopside-jadeite. Still another variety, called chloromelanite, with a strong, dark-green-to-black color, has a composition intermediate between jadeite, diopside, and acmite—a sodium iron silicate. Whether the sample is jadeite, diopside-jadeite, or chloromelanite, all of these are unquestionably identifiable as pyroxenes and are close enough to jadeite to be recognized as merely slightly variant kinds of the same species.

The crystals of jadeite are not fibrous or matted, as are those of nephrite. However, they are complexly interlocked so as to have a compact, granular appearance. Jadeite, because of this difference, is not as tough as nephrite, but at 7 it is somewhat harder on the Mohs scale. It is measurably denser, too, than nephrite with a specific gravity varying from 3.30 to 3.36. Colors of jadeite tend to parallel those of nephrite, except that a pure-white or near-white jade is almost surely jadeite, as is a vivid emerald green. Also reserved to jadeite are blue-greens, shades of lavender, and near red.

The challenge to the present-day jade carver and collector is an old one: how can jadeite and nephrite be distinguished from each other and from very similar substitutes? Most of the folklore of jade identification is worthless. For example, jade carvings do feel cool to the touch, as we are told, but so do many other stones. Simple tests can be applied, and long years of experience with jade can be used to make visual identification with a high percentage of success. In difficult cases, even among the experts, simple tests may fail miserably. Of course, there are more sophisticated methods available to the experts using expensive technical equipment and procedures that are just about foolproof.

Hardness is not a good test. Nephrite is 6 to 6½, and jadeite 6½ to 7. Then, too, there is a variety of good green serpentine, called bowenite, which looks very much like nephrite and has a hardness of 6. These hardnesses are all too close for certainty. The old trick of making a small, inconspicuous scratch with a penknife will not distinguish jadeite from nephrite, and will not eliminate bowenite or other hard species. It is useful for separating out the softer varieties of serpentine, soapstone, etc. Specific-gravity determinations are good, in that they do not require any mutilation of a carving. The entire piece can be measured. Fortunately, too, jadeite has a specific gravity of 3.3 to 3.5, and nephrite of 2.96 to 3.10. The gap is sufficient to measure and distinguish one from another. But once again, bowenite, with a specific gravity just under 3.0, confuses nephrite identifications. Visual examination sometimes helps, because the expert soon comes to recognize a certain crystallinity in jadeite that is lacking in the other stones. Nephrite and serpentine have an amorphous internal appearance. A polished surface on nephrite seems to look more oily than glassy.

Undoubtedly, the quickest and easiest of the reliable determinations is using a petrographic microscope, or a refractometer, to determine the refractive index. A trained microscopist can very rapidly differentiate between the varieties of jadeite, between jadeite and nephrite, and between jade and many of the jade-like minerals. Best of all, it takes only a few tiny grains of material, which can be removed easily from some inconspicuous place on a carving. The major drawback is that it requires an operator trained in the techniques of optical crystallography, as well as an expensive piece of equipment.

The most precise jade identification involves the X-ray diffraction powder method. This, too, requires only a very tiny amount of sample. A narrow X-ray beam is passed through the powdered mineral, which diffracts it into a pattern of lines. The pattern is recorded on photographic film, which can then be compared with a standard file of films containing patterns of known nephrite, jadeite, and other minerals. This kind of identification is fundamental because it depends on the fact that every crystalline mineral yields a characteristic pattern stemming from differences in internal structure.

A quick survey of obvious and not-so-obvious jade substitutes—many used innocently and with no intention of fraud—shows the magnitude of the identification problem. Only a very few substances are difficult to distinguish from jade. Jade-like glass, plastics, soapstone, and softer varieties of serpentine are easily separated from true jades because they are scratched readily by a good penknife. Jade carvings from ancient tombs will also scratch because their surfaces have become softened through centuries of exposure to the chemicals of decomposing bodies and soil solutions. Very old, archaic jade pieces may suffer the same alteration. If there is any doubt about these, they are worth an X-ray determination of the powdery surface material, which will show that they are still nephrite. Since jadeite was unknown in China until relatively modern times, all the ancient and altered jade will be nephrite.

Of the harder jade substitutes, the most troublesome to identify are bowenite; a green, fine-grained variety of idocrase known as californite; some fine-grained pieces of zoisite and diopside, and a few other related silicate minerals. For some, a determination of specific gravity is sufficient evidence. This, coupled with a reading of refractive index, is usually enough. Except in obvious cases, these silicate minerals are best left to trained mineralogists and gemologists for proper identification.

One of the annoying problems in jade appraisal is the practice of dyeing weak-colored or off-colored jadeite to simulate the best-quality green stones. Sometimes the dyeing is obvious. Cheaper dyes will fade in sunlight in a few days. Under magnification the dye often looks like thin, threadlike lines of color where it is trapped between jade grains but does not penetrate the grains themselves. There are dyeing processes more difficult to detect, and some so clever as to require laboratory detection. Fortunately, these pieces are not so common and are usually limited to smaller carvings. Then, too, the collector is protected somewhat in the United States by rulings of the Federal Trade Commission that such material be labeled "dyed jade."

For the buyer of jade carvings, there remains an important problem: determining the authenticity of age. Richard Gump, long familiar with the merchandising of jade carvings, gives several guidelines for this in his book on jade. Among his suggestions is this: "The style or design of an object would, generally speaking, indicate the period. Check the exact detail of design. If it is a Han vessel with a Ming or Sung design, you know something is wrong. The material used is a good indicator. If the piece is said to be pre-1784 and of jadeite, look out. If it is labeled Ch'ien Lung and of distinctive spinach-green Russian jade, which came onto the market in 1920, you would again have due cause for suspicion." The general trend of these remarks indicates the complexity of authenticating a jade carving and the long and detailed experience necessary for this kind of detective work.

Once a working knowledge of jade characteristics is acquired, another pressing question arises: what are the sources of all the rough material worked by so many different cultures in such great quantities? Until recent times it was actually believed by many scholars that all the jades of ancient Middle America came from China. They certainly couldn't have come from there, for, astounding as it seems, jade was never found in China proper. All the works of four thousand years of Chinese jade carving were executed in imported stone. There has been much speculation that the earliest carvings were done in native stone and that when these deposits ran out a thriving import trade in rough jade from other sources was begun.

225

Opposite: Fine carving of serpentine, often-used substitute which lacks several of jade's best qualities. Right: Cabochons and bracelet of jade in some of its many colors. Below left: Ch'ien Lung period (1736–95) snuff jar. Right: Modern single-piece carving of Alaskan jade in early Chinese style.

226

The evidence now very strongly indicates that the bulk of Chinese jade came from mines in eastern Turkestan, now Sinkiang Province in the west of modern China. There are references to this source in Chinese literature as far back as 200 B.C. The river valleys on the south side of the K'un-lun mountains, lying between Tibet and Turkestan, are known to have a number of jade mines that operated for at least two thousand years. Marco Polo made mention in 1272 of jade supplies moving through the nearby city of Khotan. Quantities of broken pebbles and boulders of jade also were found in the river beds draining the same mountain areas. There are other jade deposits further to the west, in the vicinity of Yarkand, and especially at the "jade mountain" within one hundred miles of Yarkand. All of this imported jade was nephrite in white, green (mostly of pale shades), and black. There was another source of nephrite in Siberia. Although close to the ancient Chinese, it was unknown to them until well into the early 1900's. These deposits are south of Lake Baikal in Siberia, and they yield a very deep spinach-green nephrite with black inclusions of chromite. Nephrite also occurs there in white or cloudy-green masses. The Russians themselves had made use of it as a decorative stone by the late 1800's, and some of it had even been carved by such great lapidaries as Carl Peter Fabergé.

The only other abundant source of good jade for the use of Chinese carvers was Burma. This jade was not discovered until the thirteenth century and took until the eighteenth to get to China, whose master carvers very likely were puzzled by its properties. It didn't work like the stone they knew. It had a different set of characteristics and very different colors. However, it was jade and highly acceptable jade, at that. The long-delayed arrival of jadeite into China had finally come about.

Jade mining in Burma has not changed much since the days of its discovery. The best deposits were found in the hills north of the Uru River. At these Tawmaw mines, just as had been done in Turkestan, fires were built on the rock face to crack the jade, which could then be split out by hammer and wedge. Quantities of jade were also recovered as boulders in loose rock deposits along and in the Uru River. Frequently this was better material, and undamaged by the mining fires. Most of the Burmese jade mined was—and is—very pale green or whitish, sometimes with rather sizeable bright green splotches. Much more rarely have the coveted solid-green pieces—in emerald and darker shades—been recovered, and even more infrequently some of lavender color.

Aside from the deposits in Turkestan and Burma, only a few other small occurrences have been noted for all the Orient. Scattered through the Pacific area there are some deposits of jade, and jade artifacts have been discovered widely scattered over the area. A deposit of nephrite exists on New Caledonia. Jadeite is found in place in the Celebes, and even chloromelanite occurs in New Guinea. The Maoris, when they arrived in their great migration to New Zealand in about the fourteenth century, quickly discovered durable nephrite in several places. Deposits they

228

Two ceremonial or dance axes of jade as illustrated in collection catalog of Emperor Kao-tsung (1127–62 A.D.). Axe on left was pale yellow and white, that on right white with green spots.

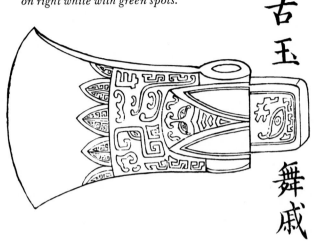

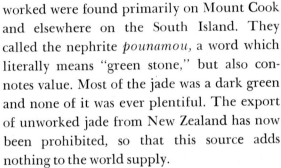

古玉
舞戚

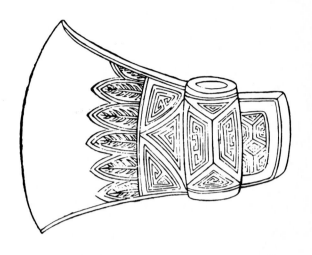

worked were found primarily on Mount Cook and elsewhere on the South Island. They called the nephrite *pounamou*, a word which literally means "green stone," but also connotes value. Most of the jade was a dark green and none of it was ever plentiful. The export of unworked jade from New Zealand has now been prohibited, so that this source adds nothing to the world supply.

A very small jadeite deposit in Japan and a more recent discovery of mostly dark-green nephrite in Taiwan complete the inventory of jade sources in the eastern part of the world. Some of the Taiwan jade has already made its way into western markets to add to the meager supply from Russia and Burma.

But what of the western world? Surely there must be a more plentiful supply there. In actual fact, a quick survey shows jade to be an uncommon mineral. As someone once said, "When scientists say jade is geologically fortuitous and the Chinese that it was a gift from Heaven, they agree in all except terminology."

It is agreed that the arts of jade carving in Central America and China reached their highest artistic levels at about the same period in history. In China, although the jade supply

THE CHINESE DYNASTIES

Prehistoric and Legendary Period	? B.C. to 1766 B.C.
Hsia Dynasty (mythical)	1766 B.C. to 1550 B.C.
Chou Dynasty	1523 B.C. to 1027 B.C.
Chin Dynasty	1027 B.C. to 256 B.C.
Han Dynasty	221 B.C. to 206 B.C.
Shang Dynasty	206 B.C. to A.D. 220
Time of the Three Kingdoms, the Six Dynasties, Western and Eastern Chin Dynasties and the Northern and Southern Dynasties	A.D. 220 to A.D. 589
Sui Dynasty	A.D. 589 to A.D. 618
T'ang Dynasty	A.D. 618 to A.D. 907
Period of the Five Dynasties, the Northern and Southern Sung Dynasties and the Chin Dynasty	A.D. 907 to A.D. 1279
Yüan Dynasty	A.D. 1260 to A.D. 1368
Ming Dynasty	A.D. 1368 to A.D. 1644
Ch'ing Dynasty	A.D. 1644 to A.D. 1912
Republic Period	A.D. 1912 —

The earliest histories do not agree on dates until 841 B.C. Before that date chronicled dates vary by 150 to 200 years. In addition, any time-scale of rulers is confused because the ebb and flow of China's fortunes brought expansion and contraction of territory, feudal states, warring kingdoms, and overlapping dynasties.

was not native, there were at least some records of possible sources and these have now been pretty well identified. All traces of American sources were lost before and during the Spanish conquest. The great mystery for centuries had been the unexplainable high development of jade carving in Central America with no obvious sources of material. The implication was that somehow jade was imported by long-lost trade routes from other parts of the world. However in 1955 the rediscovery of a jade deposit in Guatemala was reported. This jade was obviously the same kind of green diopside-jadeite used by the Mayas, Aztecs, and others. The problem was solved. Unfortunately, very little jade is added to the world's current supply from any Central American source.

Farther south, nephrite is known to have been used for tools by aborigines along the Amazon River. A small nephrite deposit has been found in place in Brazil. Old jade objects are known elsewhere in South America, but there are no obvious sources of the material other than those already mentioned.

The original natives of what is now the United States did not develop a jade industry. No artifacts have been reported in spite of the presence of significant jade deposits which have been found in recent years in the western states. By far the most important find in the United States occurred in 1942, in the countryside near Lander, Wyoming. Here the jade was collected as pebbles and boulders, some as large as the nearly 2500-pound piece in the Field Museum of Natural History in Chicago. Most of the jade had remarkably uniform texture and quality which made it ideal for carv-

ing. The color range was wide, including good green, greenish-brown, gray, and jet-black stones. Many tons were gathered, and already the best-quality material seems exhausted. Other sporadic finds of jade in California have maintained a lively interest in prospecting. Both nephrite and jadeite have been found there, but quantities are not large, nor is the quality exceptional. The beaches of Monterey County have yielded numerous pebbles and boulders of rather poor-quality nephrite, with the bulk of the recoveries being made in the late 1940's and early 1950's. Jadeite, previously known only in Burma, was discovered in San Benito County in 1936. Subsequent finds led to a jade rush into the area in 1950. These finds predated the rediscovery of Guatemalan jadeite by almost twenty years. Little jade of any value for carving has come from San Benito County, but the discoveries point out the possibility of eventually finding some jadeite at least as good as that used for the best Central American artifacts.

The great hope for a jade supply in the Americas seems to lie in western Canada and Alaska. Early explorers of the coast of Alaska, the Yukon, the country eastward to the Mackenzie River, and the coast of British Columbia came into contact with nephrite jade in general use by local Indians and Eskimos. They found large numbers of jade artifacts. The famous explorers Captain Cook and later Vancouver and La Pérouse were impressed by the native use of this incredibly tough green stone. The Alaskan Eskimos, who got most of their nephrite as boulders from the bed of the Kobuk River, flowing westward from the

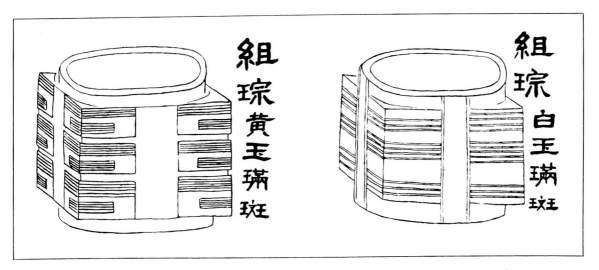

Point Barrow region, knew of the existence
of an entire mountain of this green stone along
the Kobuk River. By 1883, Lt. George T.
Emmons had penetrated the difficult area and
rediscovered "Jade Mountain," thus certifying
the Eskimo information. The entire mountain
is green, mostly from serpentine scree which
covers it, and it does contain enormous deposits
of nephrite. These Alaskan nephrites occur in
attractive shades of olive-green, yellow-green,
gray-green, and blackish-green.

As for British Columbian nephrite, the
bulk of the supply has come from the gravels
and boulders in an enormous area along the
lower reaches of the Fraser River. The deposits
are incredibly large, and each spring flood
exposes more of what seems to be an unending
supply. Sometimes British Columbian and
Alaskan nephrites resemble each other rather
closely. The British Columbia material also
occurs in several shades of green, as well as a
grayish-white with green spots. Jadeite is not
known in this part of North America.

As suggested at the beginning of this chapter, it is no accident that jade has been woven
into the fabric of early human technology
around the world because it has such unique
and utilitarian characteristics. Its sparse, sporadic, but widespread distribution has put
enough of it conveniently into the hands of
man so that he has it available to fashion for
his own purposes. Occurring with the barely
adequate supply there has been just enough
of the highest quality, best-textured, most-
translucent, beautifully colored jade to whet
the world's appetite for at least a few more centuries to come.

Sculpture in Jade

It is difficult to imagine, in this day of instant
handicrafts and assembly-line production of
art objects, that it sometimes took a lifetime
for the completion of a fine Chinese jade carving. Knowing the problems encountered in
carving even a small object, it is a staggering
thought that during the time of the great

231

Chinese jade patron, Emperor Ch'ien Lung (1736-1795), blocks up to a record size of seven tons were worked successfully. Ch'ien Lung carvings represent perhaps the highest level of skill in the long history of jade sculpting.

We do not know how long ago men discovered that they could shape jade fragments for their own purposes and that it was worth doing. The oldest jade artifacts recovered are axeheads, knife blades, and wedges—objects of a very practical nature. Early man had soon learned that jade could be honed to a fine edge and, unlike other stone tools, would keep its edge under hard usage. It was definitely worth the extra trouble to find it and shape it. What is really surprising is the early production of such tools in impractically small or large sizes, and so thinly and beautifully carved as to be almost useless for the purposes suggested by their forms. Obviously, these pieces were ceremonial or token objects never intended for use. The point is that the practical characteristics of jade which made it so satisfactory for tools also made it desirable in itself.

Jade, being such a durable material, has always been difficult to shape. In all the history of jade carving there have been few developments to facilitate the process. As with any kind of carving, the operation consists of shaping the jade by rubbing it with an abrasive. Technology has brought about the improvement in abrasives, the introduction of various improved rubbing tools, and the conversion from human to machine-powered rubbing.

Eyewitness accounts of jade working by the early Maoris of New Zealand give us some insight into methods that must have been commonly used in all early jade-carving cultures. With his *kuru* (large jade hammer) the miner broke out pieces of *pounamu* (jade). To assist in the breaking, grooves were cut by rubbing with *kiripaka* (mica schist). These grooves helped to limit breaking to predetermined directions. The final cutting and polishing were performed by long periods of rubbing with a hard stone, the *hoanga*. Apparently, even the early Maoris had stumbled on the *pirori* (rotary drill) for cutting and shaping the jade pieces. Unlike those of other primitive carvers, these drills were neither stone-tipped nor hollow; either of these special kinds of drills would have expedited the work well beyond anything possible with the simple rotating stick, sand, and water method they used.

There is very little archeological evidence of carving methods used by the ancient Chinese or Central American jade lapidaries. Of course, the carvings themselves do give some indication of the process. Since the Chinese carving technique quickly advanced well beyond anything attempted by the Maoris, they must have developed a greater variety of working tools. Also, better lubricants than water and harder abrasives were sought and tried. There is some speculation that the Chinese did not develop a drill until several centuries after they began to cut jade. Time has erased most of the evidence, but it couldn't have been too long before the drill appeared. Rather than boring simple holes, the Chinese employed the hollow reed drill, and later the hollow metal tube drill which were rotated rapidly with abrasives to cut cores in the material. With a sharp blow of a hammer the core could be broken loose, leaving a smooth hole. This method required far less effort and the removal of far less material by abrasive friction. Some drills were undoubtedly driven by rapid rotation between the palms, while ground abrasive and lubricant were fed to the working surface. This seems to have been the method also used by the Central American lapidary. The Chinese had also developed mechanized drills. Their hollow bamboo rods were driven by modifications of the same cord-and-bow arrangement used by the American Indian for his fire drills.

It was common practice for the earliest Chinese jades to be carved from slabs a half inch or less in thickness. At this distance, it is difficult to know if the slabbing of jade was due to a shortage of material, difficulty in forming thicker objects, or just because of the fashion of the times. At the start, all jade cultures performed some sort of sawing or slabbing operation. Artifacts recovered in the Fraser River area of British Columbia show conclusively that the natives there successfully used a thin, flat slab of bedded sandstone to saw through jade. The process must have been appallingly slow and wasteful of jade, but it was effective. The jade piece was first sawed partway through on one side then flipped over and sawed partway through on the opposite side. The bridge between the two cuts was then broken by a sharp blow, saving untold hours of sawing. The exposed ragged edges could then be rubbed off to smooth the slab. The Maoris, Central Americans, and Chinese certainly used similar methods. Later, the introduction of the cord saw was a definite

technical improvement. A cord drawn back and forth, as abrasive and water are fed to it, cuts hard stone surprisingly quickly. Even today the method persists, with wire taking the place of cord.

Only Chinese jade carving has continued to develop through the centuries to our own time. The art of the Central American lapidaries, although it flowered during the same period, suffered some for lack of good carving material, and then died abruptly with the arrival of the Spanish. Chinese carving development continued and eventually burst into its most frantic, ornate period in the 1700's and 1800's. Even the advent of the Chinese Communist state didn't stop it. For a period of time the craft was outlawed, but it quickly returned to favor as a way of attracting foreign currency.

Although no one knows the earliest techniques and tools for sure, rubbing-sticks of various designs, sand abrasive, and quantities of time and patience were all that the ancients seemed to use. Introduction of the rotary drill made it possible to add dimension and detail to the carvings, and to make hollow objects, such as vases and snuff bottles. These were—and are—always hollowed out before outside surface work on the carving begins. For turning the drill, the bow and the foot-treadle attachment are about the only mechanical improvements in this device in thousands of years. Since metal was introduced into Chinese technology centuries after jade had already been mastered, it had almost no real effect on methods of jade carving. Tools for sawing, drilling, and grinding became somewhat more efficient and less wasteful of jade.

Two Ch'ien Lung period carvings: 6½-inch emerald-green jadeite (below) and 14½-inch light-green jar from matched pair. Opposite: Designs of ancient dragon and hydra girdle pendants.

The earliest abrasive was probably quartz sand, which is almost always readily available. It is still widely used for many abrasive purposes. A switch from sand to crushed garnet, with its combination of superior hardness and sharp-angled grains, occurred centuries before the beginning of the Christian era. The chief gain was the time saved, but control of the shape and finish for each carved piece was reduced because of the greater cutting speed. Metal tools were introduced by the end of the Chou Dynasty, about 500 B.C. At this point the craft had matured, so that almost anything could be executed in jade.

After another fifteen or sixteen centuries, the next change in abrasive—from garnet to corundum—came about. As before, time-saving was the major gain. The final revolution in abrasives came at the very end of the 1800's, when man-made carborundum, with its very superior cutting qualities, was introduced into China. Diamond powder would be even better, but because of its cost it has had no impact on the native carving industry.

These incredibly slow changes over forty centuries have finally made it possible to produce a very intricate piece in a relatively short period of time. There are many jade connoisseurs, however, who will argue that the older carvings, done before all the changes, were really better. The hand-rubbed, meticulously cut old jades do have a smoother surface and

龍文佩
白玉瑞
辨

龍文佩
白玉黃暈

虯文佩
綠玉瑞
辨

fewer visible cutting errors, such as rough edges and irregular fine lines. It has been said that in ancient times there were no poor jade carvings. The work was so difficult and proceeded so slowly that there was more time to think about the product and very little chance of making a faulty cut which might require alteration of the carving plan. One glance at a present-day Hong Kong jade shop reveals the decline in the high art of jade carving brought about by rapid mass production with diamond abrasive and power machinery.

Whatever the methods and materials used by Chinese carvers, their purpose was clear. Jade had become to them the most suitable material in which to give artistic expression to the philosophy which was the primary driving force of their culture—and still is to a high degree. Taoism was a conception of the Universe as the result of an original cause called Tao. Development of an extensive Tao philosophy and mystique resulted eventually in a simple, nature-centered religion which worshiped the Sun, the Earth, stars, Heaven, North, South, East, West, etc. Although the objects of religious adoration were relatively few and easily understood, some of the related symbolism became so complex that its meaning is now lost and its expression in carvings as related to the basic philosophy is obscure.

The Yang-Yin symbol is one of the easiest to recognize and understand. It represents the

235

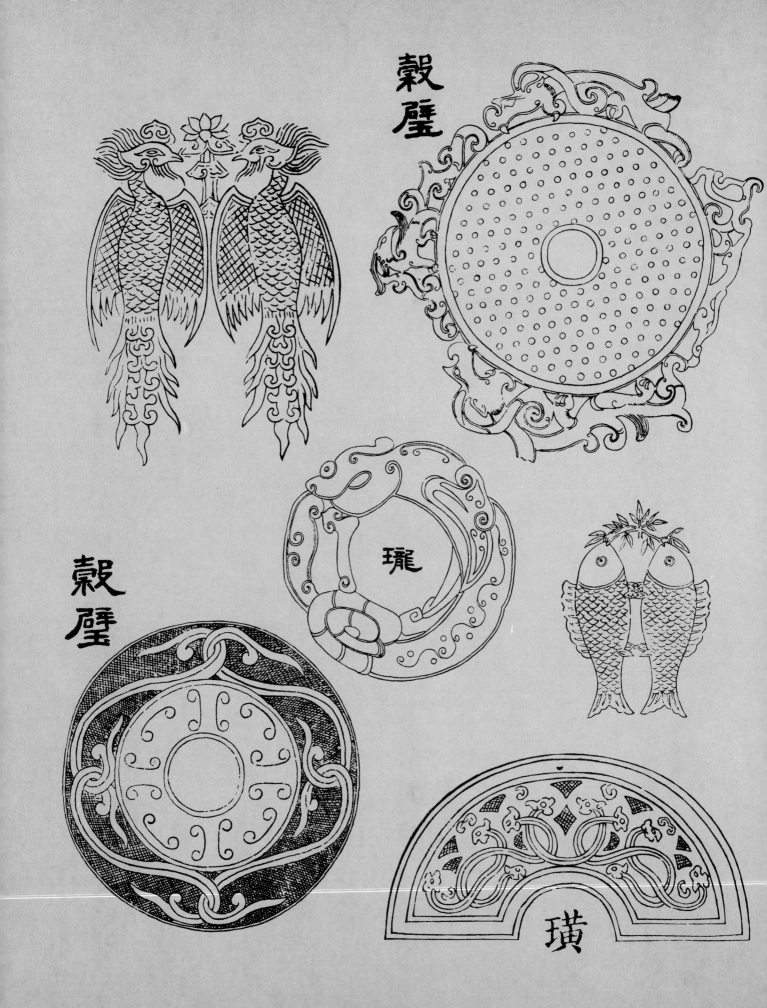

穀璧

瓏

穀璧

璜

two opposite, conflicting forces found in every action, and which are responsible for the dynamic universe. Yang is male, positive, and represented by the Sun. Yin is female, negative, and represented by the Moon. The Yang and Yin operate in the universe primarily through the agency of the five elements: Earth (Saturn), Water (Mercury), Metal (Venus), Wood (Jupiter), and Fire (Mars). These elements under the guidance of the five planets form, with the Sun and Moon, the seven rulers. Each of the elements may also be Yang or Yin, so that combinations of all these could produce broad number possibilities and astrological alternatives. Each, of course, has its symbol which can be, and often was, incised into jade. There were several other groups of symbols of great antiquity to go with these. One of the best known is the group of twelve ornaments. The origin of these is lost in time. They consist of the Sun, Moon, stars, mountain, dragon, flowery fowl, temple cups, aquatic grass, flames, grains of rice, hatchet, and symbol of distinction. Rank determined the priorities for wearing the carved ornaments, and the emperor alone had the right to wear all twelve.

Another set of designs often transmitted to posterity, inscribed in everlasting jade, were the Eight Trigrams. They make little sense now, but seem to have been designed to link the Yang-Yin principle with the five elements. The symbols consist of a series of eight symmetrical arrangements of long and short bars arranged in a circle or open square. These symbols are perhaps the most ancient of all, predating Taoism and possibly originating as far back as 2600 B.C. They have been in continuous use by jade carvers since 1100 B.C.

Two other ancient sets of figures are of primary importance in jade embellishment. These are the Ten Celestial Stems and the Twelve Terrestrial Branches. The Stems are symbols which tie the Five Elements each to something in nature and something of man. For example, wood has a Stem, or symbol, for trees and another for hewn timber. Metal has a Stem for ore and another for kettle. The Twelve Branches are a sort of Chinese zodiac. Very early, there was a division of the stars into twelve areas, each dominated by a constellation named for an animal: the rat, ox, tiger, hare, dragon, serpent, horse, goat, monkey, cock, dog, and boar. This is an all-animal zodiac, as contrasted with the partially mythical one given to us by the ancient Greeks. Combinations of the Stems and Branches are written of as being used back as far as the twenty-seventh century B.C. By the time of the Han Dynasty they had been adopted in a sixty-combination cycle of the two for the Chinese calendar. Thus the cyclical year of lightning (Stem) rat (Branch), is followed by the year of burning incense (Stem) ox (Branch)—*ping tzu* is followed by *ting ch'ou*.

As if all the symbols mentioned were not enough, China accumulated increasing numbers of them through the long centuries. Buddhism brought along the Eight Happy Omens and the Seven Gems. Taoism added the Eight Emblems of the Taoist Immortals. There are the Hundred Antiques, the Eight Precious Things, and other groups. Further, there are many individual symbols such as the swastika, which represents Buddha's doctrines or, more

237

often, Buddha's heart, and the lotus flower representing Buddhist enlightenment.

The forms of the jade carvings themselves are often as significant as the symbols inscribed on them. Perhaps best known of the ancient carvings of significant form are the Six Ritual Jades. The Book of Rites, believed to have been originally by Chou Kung about 1100 B.C., helps us to understand much about early religious practices. Translation of a passage from the book in its fourteenth-century edition by Li Ki tells us:

> With a sky-blue Pi worship is paid to
>   Heaven
> With a yellow Ts'ung to Earth
> With a green Kuei to the East
> With a red Ch'ang to the South
> With a white Hu to the West
> With a black Huang to the North.

The Pi, symbolizing Heaven, is a simple, flat, circular disk with a round hole in the middle. A Ts'ung is much more complex, but just as beautiful in form. In its simplest version it is a hollow cylinder carved in such a way that it appears to be inserted into a rectangular box of square cross section. The carving skills and dedication required to form a Ts'ung were far greater than for a simple Pi. As with most of the Six Ritual Jades, the original reasons for the particular forms are obscure, but from earliest times the Ts'ung has symbolized the earth. Later, when it became traditional to use these jades for burial purposes, the Ts'ung was placed on the chest and the Pi at the back, placing the departed squarely between heaven and earth.

The Kuei, Ch'ang, Hu, and Huang are all flat, like the Pi, but unlike the Ts'ung. Possibly derived from the shape of a primitive knife blade, the Kuei was rather plain in its early versions. A symbol of the East, or Spring, and of Imperial Power, it became customary to carve symbolic figures on its face. The Twelve Branches, The Twelve Ornaments, and any number of other potent figures have been used for such decoration. Unfortunately, the form and origin of the South, or Summer, worshiping jade is somewhat of a mystery because no authentically identified Ch'ang is known.

There has been a decided change in the Hu from its beginning form. It started as a flat, primitively formed tiger which is hardly recognizable as we think of tigers now. Strangely, it is one of the very few animal symbols dating to before the Han Dynasty. Primarily an object of worship for the West, or Autumn, the Hu has also been used as a symbol of military power or authority. This concept adds deeper meaning to the epithet "paper tiger."

Basically, the Huang is half a Pi and was for worship of the North, or Winter. For embellishment it gradually became customary to carve the curved, flat object as a dragon, or fish.

As the centuries sifted by, flowers, insects, a menagerie of animals and birds, were worked into jade. The butterfly, symbol of immortality, the bat, symbol of happiness and long life, and the three-legged toad, symbol of the unattainable, are typical. The peach was a symbol of longevity, the pomegranate a symbol of fertility, the lotus a symbol of purity, and even the lowly fungus—perhaps *Pachyma cocos*—was another symbol of longevity. Many others appeared at intervals to support and perpetu-

239

ate a vast body of mixed legend, lore, history, and religion—except that, unbelievably, none of them was a human figure.

It was more than sixteen centuries after the birth of Christ that Taoists began interpreting their abstract gods as carved human figures. Probably the first of these represented Lao-tzu, the founder of Chinese Taoism. The stimulus for this change came from Buddhism, first introduced into China from India about A.D. 67. It took almost two hundred years for the new religion to root itself, but eventually it was grafted so well to Taoism that it flourished, was totally blended into the old religion, and became uniquely Chinese in the process. Best known of the Taoist figures are the Eight Immortals, who became so popular as carving subjects during the Ming and Ch'ing Dynasties. The Eight Immortals are:

Chung-li Ch'uan, leader of the Eight Immortals who dwelt on the Mountain of Jade and became immortal through being perfectly attuned with nature. He is carved as a fat, half-naked little man with a beard, carrying a fan and a peach, or fungus, symbol of immortality.

Lu Tung-pin is carved as an older man in scholar's costume. He carries a magic sword and is revered as the protector of magicians and a doer of magical deeds.

Chang Kuo-lao, a famous magician, is often carved riding a donkey—sometimes backward—but always carrying the Yu Ku, a kind of tubular drum.

Ts'ao Kuo-chiu is the most recent of the Immortals. Usually portrayed wearing a beard, he is often dressed in beautiful robes, with a hat or cap, and is never seen without a pair of clapper-like castanets in his hand.

Han Hsiang-tzu is obviously a musician and is always seen playing his magic flute.

Ho Hsien-ku is the only true woman among the Immortals. She is credited with eating her way into the select group by consuming one of the magic peaches of immortality. A lotus flower with stem is her emblem.

Li T'ieh-kuai is always shown as a crippled beggar with a crutch. He had the misfortune of returning in spirit too late from the Celestial Regions to prevent his empty body from being destroyed. Hastily he entered the body of a dying beggar, which body became his own for the rest of his earthly life.

Lan Ts'ai-ho is an epicene figure, dressed traditionally in a blue gown, whose custom was to wander the streets singing songs about the joys of immortality. He-she (legend seems to make him male) is always depicted carrying a basket of flowers, and often with one foot resting on a spade.

Buddhism also contributed its own images, so that it became customary in Ming and Ch'ing times to carve Buddhas, Lohans, and Bodhisattvas. There are innumerable Buddhas, including the traditional Shakyamuni (Lord Buddha), Maitreya (the Laughing Buddha), and Ananda (the Teaching Buddha). The eighteen Lohans, or followers of Buddha, have been amply treated in jade, too. Best loved of all followers of Buddha are the Bodhisattvas. They are the ones who reached the peak of spiritual perfection, but spurned the final reward to stay behind and help their fellow men. The currently popular Kwan Yin was one of these. She is usually carved as a graceful, serene, gentle figure.

In a brief survey of the seemingly interminable centuries of Chinese jade carving, which produced such a wealth of aesthetically excellent and superbly carved jade objects, there is nothing visible which prepares us for the incredible flowering of the art under the reign of Emperor Ch'ien Lung during the Ching Dynasty. All of the ingredients were there, however, and it took only the proper stimulus to bring the burgeoning about. Apparently, the Emperor was a man with a highly developed appreciation of the aesthetic possibilities of jade. Personally he seemed to prefer small but beautifully carved objects in white jade, but became a patron and strong promoter of numbers of the largest, most ornate, most perfectly carved jade objects ever produced. Many excellent copies of the ancient bronze forms were produced in the royal carving shops and bore the imperial seal. The nephrite and jadeite rough material was the finest available, and the carving was meticulously done. Everything that wealth and skill could do was done to insure the quality of the final product. When Ch'ien Lung finally ended his reign in 1795, it seemed that jade carving and other typically Chinese arts entered a period of abrupt decline. The story of jade did not end, but its period of greatest glory was over. With the end of the Ching Dynasty and the fall of the Manchu princes, and with the coming of the republic in 1912, hundreds of the finest carvings disappeared from the Imperial treasuries and households. Many have since surfaced with European dealers, and have made their way into fine private and museum collections here and abroad. Large numbers of them are still in hiding and may reappear at some more propitious moment in history. However long we wait for them, it will be a short period in the long saga of Chinese jade carving.

The carving skills and techniques of the ancient Central American cultures never approached those of the Chinese. There is no question that aesthetically and artistically, the imaginative carvings of the western hemisphere have just as strong an impact as any made on the other side of the world. Ample archeological evidence points to a thriving jade-carving art among the Olmecs of the Gulf of Mexico region several centuries before the Christian era. Carbon-dating measurements push the date back to 1500 B.C. It would seem that the Chinese and Olmecs developed and perfected all the fundamental carving skills at about the same time. Many of these Olmec jade carvings have been passed down to us because they were safely buried. Among the best-known of them are the bold and impressive near-human masks and figures of jadeite. There was little attention given to anatomical detail in these very simple carvings, but the mood of a weeping child's figure, a fearsome man-jaguar mask, or a woman's face resigned to suffering, comes through very clearly. While almost everything else belonging to the Olmecs has perished, their splendid carvings assure them of a place in the history of man.

All sorts of attempts have been made to link the Central American and Chinese jade cultures by seeking out the few details that are similar. It is fairly obvious, however, that the two developed independently. There is still

Three pre-Columbian pieces illustrating wide time span of jade culture in Central America. Opposite: 5-inch Olmec figure (c. 1000 B.C.). Below: 3-inch alligator (c. 500 B.C.-300 A.D.). Right: Jade and gold Aztec pendant (c. 1450-1520 A.D.).

242

considerable confusion about the boundaries in time and space of the various Central American cultures, so that carvings and carving styles attributed to one could just as well have been passed on from another. Unfortunately, there is a tendency to associate jade with the Aztecs, because they treasured it so highly and offered it as gifts when the Spaniards arrived, a fact duly recorded by several contemporary authors. However, all the cultures—Olmec, Mayan, Toltec, Mixtec, Zapotec, and Aztec—treasured it from the beginning, until the carvings were plundered and the art destroyed by the Spanish. The Mayas probably treasured the green stone as long as any group. At the height of their production, carvers were turning out ear plugs, wristlets, anklets, and beads. Many of these objects were decorated with human and animal figures, as well as with a beautiful heiroglyphic script. By the time the Spanish arrived, the Mayan culture had already begun its decline, but there was a thriving colony of Aztec lapidaries which had transferred the art to what is now Mexico City.

Even in the beginning, good jade must have been scarce. Material authentically dated from earlier times, before the Aztecs and Toltecs, was better in color and in the uniform quality of its texture. The Spanish found only finished carvings in the possession of the natives, never any rough jade. These Aztec carvers had been very much aware of differences in color and quality. *Quetzal chalchihuitl* was the term describing precious jade which was white with a greenish tint. *Tlilavotic chalchihuitl* was green and black. Best of all was *Tolteca-iztli*, a clear, translucent green.

To all the Central Americans the magical values of jade depended more on its color and rarity than on any of its mineralogical properties. At the same time, its superior carving characteristics were recognized. It was the Aztec government which inadvertently left the strongest clues as to the sources of their jade. Each ruler in his turn prepared a tribute roll on which each town was listed, along with the taxes expected from it. Certain towns were consistently required to send jade. As the demand for jade continued to increase, the sources of supply were disappearing. Jade became so precious that larger, older carvings were sometimes hollowed in the back to remove material for new carvings.

Jade in this part of the world was put to the same general tasks—everyday use, ornamentation, and religious symbolism—as in China. Miguel Covarrubias, in his book *The Indian Art of Mexico and Central America*, describes the famous royal burial tomb at Palenque, in Chiapas, Mexico. It illuminates perfectly the religious and ornamental uses of jade: "The personage for whom the tomb was constructed, whose crumbling bones were found in an enormous, massive sarcophagus hollowed out of a single block of stone, was covered with objects of jade which gleamed in brilliant green on the layer of red cinnabar with which the corpse had been painted. On his head he wore a band garnished with large jade spangles; the locks of his hair were held in place by jade tubes; his face was covered by a magnificent mask of jade mosaic with eyes of shell and obsidian; and on his ears he wore a pair of jade earplugs incised with glyphs. His shoulders were covered with a great collar of rows of tubular jade beads, and around his neck there was a precious necklace of beads in the form of calabashes alternating with jade blossoms. His wrists were bound with long strings of jade beads forming cuffs, and on each finger he wore a jade ring, nine of them plain, one carved with the most exquisite delicacy in the shape of a little crouching man. He held a great jade ball in one hand, a square dice of jade in the other. There was a fine jade buckle or loincloth ornament, and by his feet was a jade statuette of the sun god."

Utilitarian jade was still in evidence up to the end of Aztec times. Jade wedges, spears, axes, and knives, like those made in the earliest times, remained in use right up to the sixteenth century. Central America never really moved from the stone age to the metal age. Metal drills and metal tools were never used for jade-carving chores. As a matter of fact, even the rotary drill—extensively used—never was perfected beyond the simplest twirling of a hollow bamboo reed between the hands.

With bamboo reed, wood polishers, and abrasives of quartz sand, crushed jade, garnet, and hematite, the New World carvers managed to produce a marvelous assemblage of objects. Figures of gods, such as Kinich Ahau, the Sun God; Xolotl, the horrifying guide for the dead through the underworld; and the winged serpent god were three of many. Jaguar and skull carvings were typical examples of a tendency to accentuate the terrifying and grotesque. Ceremonial axe gods, roughly in the shape of an axe blade, had the upper half carved in a figure and the lower part in a curved blade.

Bats and beaked birds, plaques and pendants, and even jade inserts and fillings for the teeth, club heads, ear ornaments, lip and nose plugs, and almost any other object imaginable were produced by the thousands. All of it perished in the sudden onslaught of a foreign culture. The subsequent poverty and prostration of all the native cultures erased a good part of what was left—even the memories.

Another Stone Age culture began to develop the use of jade at some unknown time after settlement of the two islands of New Zealand about A.D. 1000. The Maoris found jade there and promptly put it to use for work purposes. They needed good knives, hatchets, and fish hooks above all else and had little thought of making ornaments until much later. Then, these tools themselves became the first amulets. A curious and strangely contorted human figure—the Hei-tiki—with cocked head, lopsided eyes, and bowed legs, was also worn as a pendant. It and a few other carved fishhook and serpent forms are just about as far as this jade culture was able to go before the world of the European intruded with the arrival of Captain Cook in 1769. Today jade is cut in New Zealand by descendants of the same people, using power tools and carborundum drills, not the primitive tools of their ancestors.

All over the world the age of the jade cultures is finished. Now it is just another art form pursued by collectors and connoisseurs. Even so, the strong attraction of this beautiful and magic stone persists. One can judge it by the sudden scarcity of raw jade and by the escalating prices of the finished carvings of quality in the market place.

Picture Credits

Cover: Lee Boltin

Chapter 1

10–11, 14, 15: Lee Boltin. 17: *Diamant et Pierres Précieuses*, J. Rothschild, Paris, 1881. 18: Boltin. 21: German Amber Industry Guide, R. Klebs (top), *Hortus Sanitatis*, 1491. 22: Boltin (top), *Diamant et Pierres Précieuses* (bottom left), *Book of the Pearl*, G. F. Kunz & C. H. Stevenson, London, 1908 (bottom right). 24: *Gems & Precious Stones of North America*, G. F. Kunz, New York, 1890. 25: *Book of the Pearl* (left), *Hortus Sanitatis* (right). 26: Boltin/Smithsonian. 27: New York Public Library. 28: *Curious Lore of Precious Stones*, G. F. Kunz, Philadelphia, 1913. 29: New York Public Library. 31: *Mineralogie in Sachsen von Agricola bis Werner*, W. Fischer, Dresden, 1939 (top), Smithsonian Institution (bottom).

Chapter 2

34–35, 38: Boltin/Smithsonian. 43: *Grundriss der Edelsteinkunde*, P. H. Groth, Leipzig, 1887. 44: Smithsonian. 45: *Grundriss der Edelsteinkunde*. 46: Boltin/Smithsonian. 47: Boltin/Smithsonian, except top left, N. W. Ayer. 54: Boltin. 55: New York Public Library. 56: Smithsonian Institution. 57, 58: *Grundriss der Edelsteinkunde*. 59: *Minéralogie appliquée aux arts*, C. P. Brard, Paris, 1821 (top), *Grundriss der Edelsteinkunde* (bottom).

Chapter 3

62–63, 66: Boltin/Smithsonian. 67: Gemological Institute of America, N.Y. 68: Foote Prints, vol. 32, no. 1, 1960. 69: *Synthèse du Rubis*, E. Fremy, Paris, 1891. 70: Foote Prints, vol. 32, no. 1, 1960. 72: Bell Telephone Laboratories, Murray Hill, N.J. 73, 74: Gemological Institute of America. 76–77: *Synthèse du Rubis*. 79: Boltin/Smithsonian.

Chapter 4

All Boltin/Smithsonian, except page 86: Boltin/Metropolitan Museum.

Chapter 5

All Boltin/Smithsonian.

Chapter 6

142–143, 146, 148: R. W. Read. 149: *Les Pierres Précieuses*, J. Rambosson, Paris, 1870. 150: G. Switzer (top & bottom left). 154: G. Becker (top), *Les Pierres Précieuses*. 157: Smithsonian Institution. 159: *Gems & Precious Stones of North America* (top), *Jewelry, Gemcutting & Metalcraft*, W. T. Baxter, New York, 1938.

Chapter 7

162–163: Boltin/L. Slagle. 166: Boltin/Smithsonian. 169: Smithsonian Institution (top), *Precious Stones*, W. R. Cattelle, Philadelphia, 1903. 173: N. W. Ayer. 175: Boltin/Smithsonian.

Chapter 8

178–179, 183 (top & bottom left), 191 (top & bottom left), 192 (bottom): Boltin/Smithsonian. 182, 183 bottom right), 185, 187, 189, 190, 191 (bottom right), 194: Boltin. 186: *Tears of the Heliades*, W. A. Buffum, New York, 1900. 192 (top): *Mani-Mala*, R. Tagore, Calcutta, 1881. 195: N. W. Ayer.

Chapter 9

198–199: Smithsonian Institution. 202: Boltin/Smithsonian. 203: Her Majesty's Stationery Office, London. 205, 206: *Book of the Pearl*. 207: Van Cleef & Arpels, New York. 210: Smithsonian Institution. 214: Van Cleef & Arpels, New York.

Chapter 10

218–219, 222, 226–227: Boltin/Smithsonian. 229, 231, 235, 236: *Jade*, B. Laufer, Chicago, 1912. 234: Smithsonian Institution. 242–243: Boltin/André Emmerich.

Index

Picture references in italics

248